DATE DUE

DARWINIAN DETECTIVES

DARWINIAN DETECTIVES

Revealing the Natural History

of Genes and Genomes

Norman A. Johnson

OXFORD
UNIVERSITY PRESS

2007

OXFORD
UNIVERSITY PRESS

Oxford University Press, Inc., publishes works that further
Oxford University's objective of excellence
in research, scholarship, and education.

Oxford New York
Auckland Cape Town Dar es Salaam Hong Kong Karachi
Kuala Lumpur Madrid Melbourne Mexico City Nairobi
New Delhi Shanghai Taipei Toronto

With offices in
Argentina Austria Brazil Chile Czech Republic France Greece
Guatemala Hungary Italy Japan Poland Portugal Singapore
South Korea Switzerland Thailand Turkey Ukraine Vietnam

Copyright © 2007 by Oxford University Press, Inc.

Published by Oxford University Press, Inc.
198 Madison Avenue, New York, New York 10016

www.oup.com

Oxford is a registered trademark of Oxford University Press

Library of Congress Cataloging-in-Publication Data
Johnson, Norman A., 1966–
Darwinian detectives: revealing the natural history of genes
and genomes/Norman A. Johnson.
    p. ; cm.
Includes bibliographical references and index.
ISBN 978–0–19–530675–0
1. Evolutionary genetics. 2. Natural history. 3. Molecular genetics. I. Title.
[DNLM: 1. Selection (Genetics) 2. Evolution. 3. Genetics. 4. Genomics.
5. Natural History. QU 475 J68d 2007]
QH390.J64 2007
572.8'38—dc22   2006023079

9 8 7 6 5 4 3 2 1

Printed in the United States of America
on acid-free paper

Dedicated to Julie Froehlig, just for being

Imagine that the theory of evolution had never been discovered. Suppose that we are at the start of the twenty-first century without it. Biology remains based on some type of creationism, as it was before Darwin published his theory of evolution, but all of modern biotechnology is in place.

A bizarre fossil is discovered in an Ethiopian desert, bones that no one has seen before, part way between a chimpanzee and a human. The skull is small, smaller than that of any grown human, but the rest of the skeleton suggests that this man-ape monstrosity could walk upright with ease. An ayatollah declares it a creation of a Satan, a long dead minion of evil.

Sub-Saharan African children die of sickle-shaped red blood corpuscles in great numbers, but there is no contagious disease that explains the disorder, not even the endemic parasite Plasmodium that causes malaria. Molecular biologists work out that the sickling is usually due to a single amino acid change in the string of amino acids that makes up part of the hemoglobin molecule, but they have no idea why it is so frequent. An ancient curse is suspected by practitioners of voodoo.

It has been established that all known cases of inheritance are due to nucleic acids, overwhelmingly deoxyribonucleic acid (DNA), but sometimes its close chemical cousin, ribonucleic acid (RNA). The pope declares that the efficiency of these molecules shows God's Providence at work in the world.

A world where biologists only document the details of God's Creation has no mysteries. Every feature of life can be attributed to the action of this unknown, but supremely powerful, gaseous vertebrate. Yes, biologists will still work on the details, but whatever these details turn out to be, the "guilty party" is always the same: God.

Now imagine a scientific field that is devoted to finding other causes for the panoply of life, a field that invokes material causation in place of an all-powerful Creator. Then there are mysteries.

How could organs so much like video cameras have been placed on the front of our faces, giving almost all members of the species *Homo sapiens* binocular color vision?

Why is there so much variation in the amino acids making up important proteins? Why are the genes that code for proteins interrupted by noncoding DNA, of no apparent purpose? Why is more than 70% of the human genome apparently noncoding junk DNA? Why do some genes hop around, while others stay in the same place?

We can either suppose God to be perverse or look for other guilty parties. If we choose the latter alternative, further choices arise. Perhaps the causes of living diversity are secular instruments of a more subtle Creator, or the workings of a Godless universe. We don't need to decide that question right away, as biologists. But if we suppose that the living world conforms to decipherable laws, beyond those of physics and chemistry alone, though not in violation of them, then we have to work out these laws. Having worked out these lawful mechanisms, we will perforce seek out their applications to the particulars of the living world, in part to understand that living world better and in part to test these laws of life themselves.

Then and only then can we consider biologists scientists in their own right, rather than the "natural historians" of God's Creation, or the less intelligent handmaidens to chemists and physicists. It is this challenge that has been taken up by evolutionary biologists above all other biologists.

Exploring secular forces behind the machinery of life is a relatively new enterprise. For the general public, this endeavor is less than 150 years old, even though the natural history of life is one of mankind's earliest interests. And this natural history work continues at a steadily accelerating pace. Most of molecular and cellular biology is that natural history project continuing down to the level of organic chemistry, with new facts of life discovered every hour in a laboratory somewhere around the globe. But they are only creating new mysteries to be solved. For all their dazzling machinery, most cell biologists are the policemen on the beat, arresting the obvious "perps" and discovering the dead bodies, sequencing the genes and showing how the proteins fold and interact.

It is the evolutionary biologists who are the detectives, the Columbos, the Miss Marples, often older or eccentric, driving ancient vehicles and muttering to themselves about things that the cops on the molecular biology beat find baffling.

This engaging volume introduces Darwinian detective work, how we go about unraveling the guilty parties that remain behind the scenes until they are apprehended.

Many people have little interest in this enterprise. Others question its validity. Sometimes our work turns up little or nothing at first. Cases go cold, sometimes for decades. Then a new piece of evidence is turned up by a cell biology patrolman or a fossil bounty hunter, and we have a new theory explaining a mystery of life. Sometimes this theory is shot down soon after it is proposed, the victim of some ugly fact. But we work on, convinced that there has to be some underlying cause, some explanatory key that would have tickled Darwin himself, our first great evolutionary detective, and still an unrivaled master at the level of pure intuition.

But our detective work has been greatly aided by new quantitative tools, tools to which Darwin did not have access. Unbeknownst to many cursing statistics students in college, the variances and regressions that they struggle with are tools of data analysis first brought to maturity in pursuit of evolutionary mysteries. Quantitative analysis of data has grown in tandem with evolutionary biology. Now much of the evolutionary genetics that Norman Johnson describes for us is based on massive computation. (I have myself donated a successful doctoral student in evolutionary theory, Michel Krieber, to the world of stock market analysis, a forbiddingly sophisticated application of mathematics.) Modern-day evolutionary biologists can solve problems that Darwin couldn't, because our field has advanced far beyond the solid foundations that he hammered into the ground of biology; it is the most dauntingly mathematical part of biology. But not to fear. You will be spared mathematical heavy-lifting here.

There is a literal intersection of criminology and Darwinian detective work: genetic forensics. My closest colleague, Larry Mueller, serves as an expert witness in cases that involve the use of DNA from blood, semen, and other crime-scene tissues. The very fact that this technology has value in courts of law is one of the mysteries that evolutionists have been most interested in: the abundant genetic variation within our species, and most others. Without copious genetic variation, forensic DNA would be of little value. The riddle of that variation is one of the most important themes of this book.

As in many other respects at this moment in history, Western civilization has a choice between an increasingly awkward obscurantism and science. You can suppose that global warming is fanciful, that God made our bodies in His Image, that one or another ancient book contains all that a decent person should know. Or you can embrace a world that has many obscure secular causes, a world with real mysteries that take some care to figure out. Many people accept the triumphant working out of the great mysteries of physics, though there are still those who believe that the Sun revolves around the Earth, as the Bible proposes. Most people also already accept that the Earth is ancient, though few have an intuitive sense of its enormous age. It is now time for reasonably informed people to add biological evolution to their repertoire of general knowledge.

Norman Johnson introduces the educated reader, one who knows perhaps a smattering of science, to some of the most exciting mysteries in biology, mysteries that only evolutionary biologists have the hubris to try to solve. Like the cosmologists of physics or the theorists of quantum mechanics, evolutionary biologists delight in paradoxes and puzzles, organisms that shouldn't exist, and genes that seem to defy reason. Though it doesn't figure in this slim book, I have devoted much of my life to the mystery of aging. The fact of aging has been known for millennia. The difficult questions have revolved around its origins, and whether there is any prospect of slowing its onslaught. To my great pleasure, the culprits behind this crime turn out to be both quite Darwinian and quite guilty. I am hot on their tail, and relishing the chase.

Norman Johnson is now going to entice you to learn about evolutionary detective work, its twists and turns, surprises and satisfactions. I know you'll enjoy it, too.

Michael R. Rose
California, July 2006

Preface

As a child of the 1970s with dim memories of the Apollo space program, I have sometimes felt that my childhood expectations of the future have not been fulfilled. Where are the jetpacks? Where are the lunar colonies? Why haven't we been to Mars or even back to the Moon?

Two major technological transformations of the 1990s, however, have clearly exceeded my expectations. The first transformation is an obvious one: the Internet. The year 1991 saw the development of the hypertext transfer protocol (HTTP) by Tim Berners-Lee, a technique that made the rapid explosion of the World Wide Web possible.[1] By the end of the decade, the Web contained billions of pages; indeed, the search engine Google claimed in the year 2000 that it had indexed 1.3 billion of them.[2] In the years since the turn of the millennium, the growth of the Internet has accelerated and expanded into new areas of multimedia, such as podcasting and videoconferencing. Today, the Internet and other, associated computer technologies have dramatically changed how we work and how we approach many other aspects of our daily lives.

Less heralded is another revolution: a boon in genetics and biotechnology that has changed the landscape of biology and medicine. In large part, this revolution has come from great dramatic leaps in the ability of scientists to ascertain the genetic information present in the sequences of DNA molecules. Although we have understood that DNA is the genetic material since the 1940s and have known DNA's biochemical structure since 1953, scientists had not had the ability to rapidly determine the sequences of DNA until much more recently. Moreover, much as computational speed and power has exploded, so also has the ability to sequence DNA.

Our sequencing capacity is such that we no longer just sequence individual genes one at a time; now we can sequence the totality of DNA information present in individual organisms, an entity that geneticists call the genome. Coinciding with the fiftieth anniversary of Watson and Crick's unraveling of the structure of DNA, the DNA information contained in the human genome was fully ascertained in April 2003. This was no small feat, given the size of the human genome (just over three billion of the basic units of DNA, which biologists call nucleotides or base pairs). Yet the human genome was sequenced—sooner than expected, and under budget.[3] In addition to the human genome, we now also possess knowledge of the complete sequences from the genomes of over five hundred organisms. Hundreds more will be completed within the next decade.

The pace of discovery is also accelerating due to advances in the technologies used to sequence genes, as well as the development of new and more powerful computers and computational tools. These advances led to the emergence of the interconnected fields of genomics (study of genomes), bioinformatics (the intersection of biology and computational sciences), and proteomics (the study of the proteins that are produced from the genes in genomes and their interactions). Many biologists now perform most of their research not at the lab bench or in the field but at the computer, analyzing the information of genes and genomes already contained in ever-expanding databases.

Yet tension exists among different communities of biologists. In contrast with the rapidly accelerating pace of research in genomics and other new high-tech fields, older, more-established disciplines such as conservation biology, ecology, and evolution receive far less attention and funding for research. Geneticists and molecular biologists often dismiss these older fields with the pejorative label "natural history." When challenged, those studying ecology and evolution often counter with statements that the lab-bound scientists aren't really studying nature. Such tensions are not new; indeed, they predate the rise of molecular genetics, starting in the middle decades of the last century.

Natural history should not be a pejorative label. Usually defined as the study of entities (including organisms) in their natural environments, natural history includes studies of their origins, evolution, interrelationships, and behavior. Genes and genomes—the provinces of the molecular biologists— also have natural histories. In fact, a great deal of molecular biology over the past half century is about characterizing the genomes of organisms, a topic that might be considered descriptive natural history. And researchers found quite a few surprises! It is striking that only 1 to 10% of the genomes of animals actually codes for genes. Although vertebrates (animals with backbones) generally have larger genomes than do insects and other invertebrates, the vertebrate genome is less gene rich. In fact, although the human genome contains about ten times as much DNA as the genomes of roundworms and fruit flies, the number of genes in humans is only about twice the number of genes in these supposedly "simpler" organisms. Why is this? And what does the 90 to 99% of the genome that does not encode for genes do? These questions have been only somewhat answered, and will require a deep understanding of molecular genetics as well as evolutionary biology to be fully addressed.

Not all genes within the same genome have the same evolutionary histories; different regions of genome can experience considerably dissimilar environments. The way we inherit our genomes facilitates dissimilar evolutionary histories for various parts of the genome. When gametes are made, the genome—even including genes that are on the same chromosome—is scrambled in a process known as genetic recombination. As we will see in chapter 6, the time and place identified for the most recent common ancestor for all humanity depend upon which gene is chosen for study.

This book explores how evolutionary biologists use the tools of molecular genetics to unravel the secrets of the natural histories of genes and genomes. Much like detectives looking to ascertain the circumstances behind a crime, evolutionary biologists can make and test inferences about the nature of the evolutionary pressures that have shaped the organisms that harbor such genes. The clues they obtain come from examination of patterns in the information contained in genes. By immersing themselves in the natural histories of genes and genomes and by using the tools of mathematical evolutionary theory, such Darwinian detectives can and are asking questions such as these:

Why do some organisms have a much lower genome size than their close relatives?
Has natural selection, and if so what kind of natural selection, operated on a particular gene?
What were the genetic changes that were associated with our becoming human?

In addition to telling stories about genes and the organisms that harbor them, I also aim to give a sense of the thought processes scientists use to unravel these puzzles of the natural world. Although science is based on facts, it is much more than just facts. Ultimately, science is about making sense of the universe and its parts. Science involves forming explanations for the patterns we observe; in this regard, it is a creative activity. These explanations, formally known as *hypotheses*, are tentative at first. In addition to formulating hypotheses, science requires the rigorous testing of those hypotheses with experimentation and further observation. Unfortunately, most nonscientists (even those with a strong interest in science) lack a clear understanding of exactly how science is practiced today. A major goal of this book is to facilitate the public's understanding of the professional science of evolutionary biology, and in doing so, maintain clarity but with as little dilution of rigor as possible.

The general public's lack of understanding of science, and of evolution in particular, has enabled the rise of a most pernicious political movement: a new faith-based creationism that presents itself as if it were actually science. Between 1999 and 2006, school boards and legislatures in several American states have attempted to either minimize the teaching of evolution or to introduce a form of creationism called "Intelligent Design" as an alternative to the teaching of Darwinian evolution in public schools. Countering these efforts—which are entirely political and religious, and not scientific—will require more education of the general public about what evolution is and how science is practiced.

The book begins with a chapter demonstrating why evolution not only is the central organizing principle of biology but also is worthy of study for very practical, real-world reasons. In the second chapter, we will also examine why Intelligent Design creationism is not science, and why it is not an alternative to evolution for explaining the natural world. Here we will also encounter the Discovery Institute, the Seattle-based think-tank that has been supporting and funding the Intelligent Design movement. The mission of the Discovery Institute extends beyond promotion of Intelligent Design; indeed, its aims include undermining and overturning the principles of the Enlightenment on which the United States was founded and replacing these principles with theistic ones.

Next we'll turn to Darwin's prime explainer of evolutionary patterns: natural selection. Consider the honeybees; they have evolved elaborate "dance" behaviors that allow them to communicate to others in their nest the location of food sources that may be hundreds of yards away. Natural selection is a requirement for the evolution of these dance behaviors and other intricate adaptations to the environment. The process of evolution, however, is more complicated than just adaptation via natural selection. First, natural selection is not the only evolutionary force; random processes such as the sampling error known as genetic drift also influence evolution. Moreover, natural selection comes in several forms. The type of selection that increases the frequencies of advantageous genetic variants and ultimately allows for the bee dance and other adaptations is often called positive selection. In contrast, selection against recurrent deleterious mutations is called negative selection. Selection can also actively maintain different variants of genes; this is called balancing selection.

In chapters 3 through 5, we'll explore how evolutionary geneticists go about their detective work of finding the footprints of selection on genes. Chapter 3 begins with a discussion of the importance of negative selection in the evolution of DNA and protein sequences; at the molecular level, the weeding out of deleterious mutations by negative selection is usually more frequent than the actions of either positive or balancing selection. After discussing a negative selection, we turn to a conceptual framework that explains the patterns of molecular evolution that would be expected if the fates of genetic variants were determined primarily by mutation, genetic drift, and negative selection alone. This conceptual framework, by the Japanese geneticist Motoo Kimura, is called the neutral theory of molecular evolution, and it forms the foundation of many of the studies that are explored in the remainder of the book.

Evolutionary geneticists now use Kimura's neutral theory of molecular evolution to establish tests for positive and balancing selection. These Darwinian detectives start with figuring out what would happen if no positive or balancing selection operated. They then look for deviations from those expected patterns, and these deviations (if found) can provide support for claims of positive or balancing selection. Chapter 4 introduces some of the basic tests used to detect positive natural selection, and presents biological examples in which positive selection has been detected. Next, chapter 5 describes the tests used for detecting balancing selection. Many of the best-known cases of balancing selection involve human diseases such as malaria.

The discussion of how evolutionary geneticists detect different forms of natural selection prepares us well for the subject of the next section: what genetic sequence information can tell us about human origins and evolutionary history. In chapters 6 and 7, we'll look at the tracking of genetic changes to unravel details about our recent evolutionary history. One surprising finding is that the common ancestor of all human females lived substantially earlier than did the common ancestor of all human males. Answers from these studies allow us to address whether Neanderthals contributed to our gene pool, and whether Neanderthals and modern humans should be considered two separate species or two subspecies of the same species. Chapter 8 explores the data that show that our closest relatives are the common chimpanzee and the bonobo (formerly known as the pygmy chimpanzee).

What were the genetic changes that made us human? Perhaps we will never know the answer to that question, but we can address a related question: which genetic changes that occurred along the human branch of the evolutionary tree were also driven by positive selection? chapter 9 tackles this question, building on the tests discussed in chapter 4 with new information from the human and chimpanzee genomes. In chapters 10 and 11, we return to more recent human evolutionary history, asking questions about correlations between linguistic and genetic variation, and the genetic changes that occurred in the plants and animals that humans domesticated.

In the final chapter, we turn to the properties of genomes and, in particular, the variation in the size of genomes in different organisms. For example, why do pufferfish have much smaller genomes than do their close relatives? The chapter also explores a recent hypothesis that many general features of the genomes of different organisms may be the result of a few key parameters such as the population size and the mutation rate.

In writing this book I have tried to accurately present to a wide audience a glimpse into how scientists solve complex problems. Although both evolutionary biology and genetics rest upon a rich body of mathematical theory, I have kept the mathematics to a bare minimum and have instead presented concepts in verbal terms. In chapters 3 and 4, the more mathematical material is set aside in boxes from the rest of the text. Occasionally, I have simplified concepts and results; in some cases, I provide more information in the end-notes, and in others, I provide references to those who wish to delve deeper into the literature. I have also ventured into areas, such as anthropology and linguistics, in which I am not an expert. To practitioners in these subject areas, I apologize for any misrepresentations or oversimplifications. I have minimized but not completely eliminated jargon. To assist with understanding vocabulary that may be unfamiliar to the general reader, I have introduced special terms upon first appearance and have provided a glossary at the back of the book.

Acknowledgments

It is difficult to imagine publication of *Darwinian Detectives* coming to pass without the enthusiastic guidance and assistance of Peter Prescott, my editor at Oxford. Since the time I pitched the idea for such a project to him back in summer of 2004, Peter has been a steady champion. I am also grateful to Alycia Somers, Keith Faivre, Suzanne Copenhagen, and the rest of the staff at Oxford University Press for their excellent work.

To my family, friends, and colleagues: thank you for your support and your continued attention. I thank my mentors and teachers, espe cially Bruce Grant. I am also grateful for the support given by the Department of Plant, Soil, and Insect Sciences and the Graduate Program of Organismic and Evolutionary Biology at the University of Massachusetts at Amherst.

Alan Dickman at the University of Oregon read the entire first draft. His insightful comments and suggestions improved the book substantially. I thank the following for reading and commenting on one or more chapters from the book: Carol Boggs, Julie Froehlig, Jay Hegde, Chad Hoefler, Olivia Judson, Michael Lynch, John McDonald, Mohamed Noor, Ben Normarck, Sarah Purvis, Lynnette Leidy Sievert, Michael Wade, and Sean Werle. Several anonymous reviewers also read parts of the manuscript at various stages; their assistance was very valuable. I thank Sylvia Purvis, research librarian at Central Connecticut State University (and my aunt), for tracking down the source of an obscure quote.

Sean Werle assisted with the construction of many of the figures. I am deeply grateful for his help. I also thank Kara Belinsky, Ray Coppinger, Simon Fisher, Laurie Godfrey, Chad Hoefler, Sarah Huber, Sharlene Santa, and Frank Williams for assistance in providing illustrations.

Contents

DARWINIAN DETECTIVES

1

The Baby with the Baboon Heart
Why Evolution Still Matters

*Medicine is magical and magical is art. / Think of the boy in
the bubble and the baby with the baboon heart*

—Paul Simon, "The Boy in the Bubble"

*Whether it's preventing a flu pandemic or tackling malaria,
we can use our knowledge of evolutionary processes in
powerful and practical ways, potentially saving the lives of tens
of millions of people. So let's not strip evolution from the
textbooks, or banish it from the class, or replace it with
ideologies born of wishful thinking. If we do, we might find
ourselves facing the consequences of natural selection.*

—Olivia Judson (Judson, 2005)

■ Baby Fae

On October 14, 1984, a baby girl, known as Baby Fae, was born prematurely with the left side of her heart not fully developed. Given this condition, hypoplastic left-heart syndrome, her chances for survival were very poor; desperate measures were certainly called for. Twelve days later, Dr. Leonard Bailey and his team at Loma Linda University Medical Center (a Seventh Day Adventist institution in California) implanted a heart from a baboon into Baby Fae.

Baby Fae survived three weeks after the transplant but died on November 15, 1984. Dr. Bailey described the cause of her death as "complications that caused her red blood cells to clump together, obstructing microcirculation throughout her body."[1] Other sources noted that Fae's death was the result of her immune system attacking the foreign tissue that was the baboon heart. Indeed, immunorejections often plague cross-species transplants. Bailey and his staff did use antibiotics and other measures to minimize the risk of rejection, but their measures did not prevent the immune reaction.

Why use a baboon heart? Other primates, such as gorillas and chimpanzees, are considerably more closely related to us, with respect to evolution. Indeed, this difference in closeness of relationship is reflected in taxonomy; humans, chimpanzees, and gorillas are apes while baboons are monkeys. When asked why he had used a baboon and not picked a donor from an animal more closely related to humans, however, Bailey said that he didn't believe in evolution. Bailey went on to say:

> The scientists that are keen on the evolutionary concept that we actually developed serially from subhuman primates to humans, with mitochondrial DNA dating and that sort of thing, the differences have to do with millions of years. That boggles my mind somehow. I don't understand it well, and I'm not sure that it means a great deal in terms of tissue homology.[2]

It is ironic that Dr. Bailey used the term "tissue homology"; the word *homology* means "similarity due to shared ancestry."[3] Although we share ancestry with the baboon, we share much more recent ancestry with chimpanzees and gorillas than we do with the baboon. For that reason, we are more similar evolutionarily to chimpanzees and gorillas. Tissue homology does depend on overall homology and evolutionary divergence since the common ancestor.

We cannot, of course, be positive that Baby Fae's operation would have been successful even if an animal more closely related to us were used in the transplant. So many possible things could have gone wrong. Yet the cavalier dismissal of evolution—the central organizing principle of biology—by a medical doctor is deeply troubling when lives are at stake. Despite its appearances and Paul Simon's lyrics, modern medicine isn't magical; instead, it is based on scientific principles that include evolution.

■ Avian Flu Looming

On November 1, 2005, President George W. Bush unveiled his administration's plan to respond to the threat of a bird flu pandemic (a widespread epidemic). The response included a request for $7.1 billion of emergency funding for the production of vaccines, the stockpiling of anti-viral drugs, and surveillance measures to respond to outbreaks.[4] At the time of Bush's announcement, only 122 people had contacted this particular avian flu, the H5N1 influenza strain that first appeared in Hong Kong in 1997. And yet for those who acquired the virus from birds (mainly chickens), the flu has been deadly, as just over half of those infected died.

In the months following announcement of the Bush plan, avian flu continued to spread both in bird and human populations. By early 2006, infected birds were collected from Asia, Africa, and Europe; many biologists and public health officials suspect it may reach birds in the Americas soon. Human deaths from infections have occurred in Egypt, Turkey, and Iraq—countries far removed from Southeast Asia, the source of this flu strain. In six months after the first report from the Bush White House, 84 additional human cases of avian flu were recorded, about as many as had been recorded in the two prior years. Coinciding with the six-month anniversary of their first report, the White House released a second document with further details about how the federal government would respond if avian flu were to reach the United States. Because so many localities could be hit with the flu at one time, the report suggests that the federal government may not be able to respond to this crisis as well it does to hurricanes and other natural disasters. Were human-to-human transmission of this flu to reach the United States, domestic travel could be restricted, schools and workplaces closed, and quarantines imposed. Among other recommendations, this report advises people to maintain a distance of three feet from each other in the event of the pandemic.[5]

Bird flu has not yet reached pandemic stage, and it may never reach it. As of May 2006, avian flu in humans remains in what the World Health Organization denotes as "self limiting" or Phase 3; the virus cannot spread from person to person (except possibly in rare cases of very close contact). Even in the absence of intervention, the virus would not spread among humans. So why the concern?

The problem is that viruses can adapt to novel environments quickly; through natural selection and other evolutionary processes, they acquire new characteristics. Although the current H5N1 bird flu virus does not transmit easily to humans, just a couple of small genetic changes could lead to the flu being transmissible from human to human. Why is it that we don't get bird flu easily?

Part of the reason that humans don't acquire and transmit bird flu more easily is that just getting into our cells is an obstacle for this H5N1 flu strain. Flu viruses carry a molecule called hemagglutinin that binds to receptors on the host cell, enabling the virus to enter the cell; the H of H5N1 denotes that this virus strain has the fifth major variant of hemagglutinin. Various molecules known as sialic acids are on the surfaces of our cells; these molecules play a number of roles in a variety of systems as diverse as the immune system and neurobiology. The H5N1 hemagglutinin binds easily to the 2,3 form of sialic acid, but usually not to the 2,6 form. Fortunately for us, the surfaces of cells in the human trachea contain mainly the 2,6 form; cells lower in the respiratory tract, however, are coated with the 2,3 form. The ease at which the virus can adapt to being able to bind to the 2,6 form of sialic acids is an important aspect of the more general question of the likelihood of humans contacting H5N1 more readily. Unfortunately, it appears that H5N1 may already be evolving in this manner; the virus isolated from a human in Turkey early in 2006 is a mixed population, with some viral particles showing affinity to binding to the 2,6 form.[6]

Concerns about "bird flu" are not new; indeed, for years many public health experts have been fretting about the possibility of a deadly flu virus acquiring human-to-human transmission. Such a change in the virus could lead to a pandemic that would infect millions. A *New Yorker* article written in early 2005 called the avian flu "nature's bioterrorist."[7] In the worst-case scenario, a pandemic of this flu could kill over 100 million people worldwide.

Nobody knows what are the chances of such a worst-case scenario, but many experts, including Dr. Anthony Fauci, the director of the National Institute of Allergy and Infectious Diseases, say "It's a matter of when, not if" a flu pandemic will strike.[8]

We have experienced flu pandemics before in the not-so-distant past. Between 1918 and 1920, influenza killed as many as 50 million people worldwide. This 1918 influenza pandemic, known sometimes as the Spanish flu, caused the deaths of far more people than did World War I. An estimated 700,000 Americans died from the Spanish flu, no one knows exactly how many; in contrast to the annual flu that kills mainly the elderly, most of the deaths from the 1918 flu were of young to middle-aged adults. The toll from the 1918 flu was so great that the life expectancy in the United States was briefly depressed by about ten years. Three weeks before Bush's plan was unveiled, researchers published in *Nature* the genome of the virus that caused the 1918 pandemic. These data demonstrate that the virus is a bird flu that adapted to humans via one or a few genetic changes.[9]

The 1918 flu was not the only deadly flu that we faced in the twentieth century. In 1957 and 1968, milder influenza epidemics each killed roughly a million people. Unlike the 1918 flu virus, the viruses of these latter influenzas were recombinants—genetic mixtures—between a bird virus and a human virus.

How do we know the origins of these flu viruses? Through the use of tools developed by evolutionary biologists to examine the evolutionary relationships of organisms. These same techniques are used to examine the origins of other pathogens, such as the SARS virus responsible for the outbreak in 2003 and the anthrax bacteria used in the bioterrorist attack in the fall of 2001.

Both the bird flu and the "ordinary" flu viruses evolve rapidly. Their short generation time enables them to go through many generations within a short period, and the faster turnover of generations allows for a rapid rate of evolution.[10] Flu viruses also evolve rapidly because their mutation rate is high. Their high mutation rate and short generation time, together with their enormous population size and other factors, make flu viruses highly evolvable. Because of their rapid evolution, new flu vaccines are needed each year; old vaccines are less than adequately effective against the new flu. The principles of evolutionary genetics are used to predict what the next year's flu will look like, and potentially to ascertain whether a radically different and thus more deadly virus is on its way.

Understanding how the bird flu virus can and will adapt to humans requires knowledge not just about the virus but also about the principles of evolution and population genetics. As Thijs Kuiken and his colleagues in a review on bird flu adaptation note, "the effective rate of virus adaptation is not simply determined by the overall rate at which mutations arise, but by the fitness of these mutations, particularly the proportion that are advantageous in multiple hosts."[11] We will discuss in the future (mainly in chapter 4) how biologists can detect the action of positive selection, the form of selection that yields adaptations.

■ Evolution as an Enemy, or Resistance Is Not Futile

According to some estimates, about 100,000 people die each year from infections acquired in hospitals, a dramatic increase from 1992, when only 13,300 people died from such infections. This serious health crisis has been brought on due to the rapid evolution of resistance of bacteria to antibiotics. Consider the antibiotic vancomycin. Among other purposes, it has been used to treat urinary tract and other bacterial infections caused by *Enterococcus*, spherical bacteria that typically live in digestive and genital tracts and are a major source of infection. Prior to 1989, there had been no reported cases of *Enterococcus* that were resistant to vancomycin. Four years later, 8% of hospital *Enterococcus* infections were resistant to the antibiotic. In the years since, the proportion of resistant *Enterococcus* has continued to climb.[12]

Bacteria rapidly evolve resistance to the agents used to control them. In this sense, evolution is an enemy that exacts a heavy toll on our society. The increasing resistance of several disease-causing bacteria to many antibiotics is a serious and escalating public health crisis. Many species of these pathogenic bacteria are resistant to multiple antibiotics, including some that had previously been antibiotics of "last resort." The evolution of antibiotic resistance is often so rapid that new antibiotics are virtually ineffective a decade or less after their introduction. This resistance has resulted in tens of thousands of deaths every year in the United States alone. In addition to the human toll, bacterial resistance to antibiotics costs Americans tens of billions of dollars each year in extra medical care, increased duration of hospital stays, and lost productivity.[13]

Over one million people in the United States and nearly 50 million worldwide are infected with HIV, the virus that causes AIDS. Over two million people die from HIV each year in Africa alone. Part of the reason that HIV is so deadly is that the virus evolves so rapidly that the immune system of the infected person cannot keep up with it. Like the flu virus, HIV's rapid evolution arises from its high mutation rate. This rapid evolution also makes it unlikely that application of a single drug will effectively halt the progression of AIDS in infected persons. Instead, a cocktail of several drugs operating on different parts of the life cycle of the virus is the most effective treatment. However, even these "wonder drug cocktails" are becoming less effective due to the evolution of the virus. In his book *The Evolution Explosion*, Stanford biologist Steve Palumbi explains how we can delay the evolution of drug resistance in HIV-infected individuals by applying evolutionary principles.[14]

Farmers spray their crops with massive amounts of various insecticides to reduce damage from insects. Such large amounts are thought to be required because the insect pests, owing to their short generation time, can quickly evolve resistance to each chemical used to control them. This evolution of resistance costs the United States approximately $2 billion each year and puts our food supply in jeopardy. Furthermore, the large amounts of chemicals used pose a direct danger to human health. Farmers often court failure by *intensifying* their attacks on the pest species by using more pesticides or targeting specific pest species. The principles of evolution tell us that this is bound to intensify the evolution of specific resistance, and that a better strategy from an evolutionary point of view is to *diversify* the attack to prevent specialized adaptations against the pesticide. This is the basis of many "new" pest-control techniques, which are based on evolutionary principles—even if their users do not recognize the fact.

The application of evolutionary principles has already helped in the fight against the evolution of resistance, but the evolution of resistance is a relentless enemy. Countering it will require the training of new generations of scientists, doctors, and public health officials in evolutionary principles. In addition, combating this "war against resistance" will require a more informed public.

■ Plant and Animal Breeding

Darwin himself was an avid breeder of pigeons and other animals and plants, and in some ways, with its descriptions of numerous varieties of pigeons, the first chapter of his *The Origin of Species* reads more like a breeder's handbook than it does a polemic. In fact, Darwin used the artificial selection that occurs in breeding as a metaphor for the natural selection that is the major (but not exclusive) force in his theory of evolution.

Countless biologists since Darwin have had research interests in practical applications of evolutionary principles in the breeding of animals and plants. These biologists developed the field of quantitative genetics to deal with understanding how selection and other forces will affect traits, such as milk yield in cows or oil content in corn, that are due to multiple genetic and environmental factors. The genetic basis of agriculturally important traits often complex; moreover, different traits may be affected by the same genetic changes. For this reason, traits are often correlated. To take a hypothetical but reasonable example, a desirable trait in cattle such as lean muscle mass may be correlated with a less desirable trait such as fat content. Selection for increases in the first trait may lead to undesirable increases in the other. So, how does one get larger but leaner cows? It is problems like these that evolutionary quantitative geneticists can solve.

Evolutionary genetic principles are also important in addressing concerns in conservation biology. Populations that are maintained at low numbers for too long will lose genetic variability; for this reason, they will be less able to respond to new challenges, such as diseases. For example, cheetahs are almost completely genetically uniform, and most cheetah males have abnormalities in their sperm. The principles of evolutionary genetics inform conservation policies by determining guidelines for minimum population sizes needed for the maintenance of genetic variation, as well as breeding management strategies.

■ Molecular Evolution

Molecular biology provides arguably the most convincing evidence that all known life is related—that is, shares common descent. The most striking feature that demonstrates common descent is the commonality of nucleic acid DNA as the genetic material across all known life. Some viruses (HIV, for example), known as the retroviruses, use the related nucleic acid RNA as their genetic material, but even these retroviruses will convert their RNA information into DNA information at specified times. Moreover, all other forms of life make use of RNA. Not only is all life based on the nucleic acids DNA and RNA, but also—aside from a few minor exceptions—all life uses the same genetic code to convert the information in the nucleic acids into amino acid information in proteins. In addition, basically the same machinery, with a few modifications, is used in cellular machinery in humans, fish, insects, flatworms, plants, and bacteria.

The information stored in a strand of DNA is its sequence of nucleotides. At any particular site along a strand of DNA, there can be any one of four nucleotides: adenine (A), cytosine (C), guanine (G), or thymine (T). The DNA code is simply the order of these Gs, As, Ts, and Cs. If you were to line up the DNA sequence of the rat gene that encodes for the hemoglobin protein (which carries oxygen in the bloodstream) and its human counterpart, you would see stretches where the sequences are the same interspersed with sites that differ between the two. Because information in proteins is encoded from DNA information, the proteins made from human and rat genes are also slightly different. In more distantly related organisms—say, humans and frogs—the divergence in the gene counterparts is generally greater.

The textbook definition of evolution is simply the change of the frequencies of gene variants in populations over time. Although both this definition and the vernacular definition of evolution imply change over time, ironically evolution at the molecular level is often a very conservative process. About 99% of human genes have a counterpart in mice, despite the 80 to 100 million years of evolutionary divergence since humans and rodents last shared a common ancestor.[15] Progressively more distantly related organisms—kangaroos, turtles, sea urchins, and flies—share fewer and fewer genes with us. Still, researchers continue to find a surprisingly large number of plant counterparts of human genes.

R: ATG GTG CAC CTG ACT GAT GCT GAG AAG GCT GCT GTT AAT
H: ATG GTG CAC CTG ACT CCT GAG GAG AAG TCT GCC GTT ACT

Figure 1.1
Comparison of DNA sequences. The first 39 nucleotide sequences of the beta-chain of hemoglobin from human (H) and rat (R) are presented. Nucleotide differences are highlighted in gray.

That we share 99% of our genes with mice doesn't mean that humans and mice are 99% genetically identical; in fact, we are a good deal less than 99% identical to mice, because each of the gene counterparts we share with them have diverged. In other words, humans and mice have a large proportion of genes arising because humans and mice share a common ancestor, but in the course of time through the action of evolutionary forces, humans and mice now have different variants of the same genes.

A consequence of the divergence of genes with increasing evolutionary distance is that genes from one species do not function as well in the body of another. Moreover, that function declines as the foreign genes are taken from progressively more distant species. The malfunction occurs because genes interact with each other; the foreign genes have evolved separately from the resident genes and, hence, do not interact as well with them. One example concerns the mitochondria, an organelle within cells that contains its own genes. These mitochondrial genes interact with genes from the cell nucleus to provide the machinery for oxygen-driven (aerobic) respiration. In cell cultures, human cells wherein the human mitochondrial genes were knocked out and replaced with those of gorilla or chimp (our closest relatives) can still function.[16] In the same experiment, but with mitochondrial genes taken from more distant relatives—orangutans and monkeys—the cells died because they could not utilize oxygen properly in respiration. We will return to the mitochondria in chapter 6, when we discuss how mitochondrial DNA has been used to trace human ancestry.

Molecular evolutionists find that different genes evolve at different rates; some genes (and thus their proteins) have changed a great deal between humans and mice, whereas others are nearly, or even completely, identical. The differences in the rates of evolution among genes reflect differences in the intensities of evolutionary forces, including mutation and selection, that have operated on these genes. A major theme of this book is how evolutionary biologists can draw inferences from observed patterns in the rates of evolution across different genes.

It is at the molecular level, especially, where we see the connection between microevolution, the evolution that takes place within populations and closely related species, and macroevolution, evolution at higher levels of biological organization and deeper timescales. Although there may (or may not) be differences in the importance of the various evolutionary forces operating at each of these levels, macroevolution and microevolution are still the result of underlying changes in DNA.

■ Patterns of Variation

We've spent billions of dollars on the Human Genome Project with the ostensible purpose of benefiting for human health. Although it has been useful in uncovering numerous new genes, the completion of the Human Genome

Project is in some ways just the start; the real payoff to the biomedical community has only just begun.

The completion of the genome sequence—that is, knowing the sequence of all three billion or so nucleotides—sets up a framework to study genetic variation among humans at the scale of the whole genome. Such studies of variation will be instrumental in determining why some people have a greater propensity for acquiring certain diseases that have a genetic basis (run in families), but one that is complicated.

Let's consider genetic variation. Even though, on average, two people differ from one another at only about 1 in 1,000 nucleotides, the vast size of the genome means that millions of different variants are within the human population. Most of these variants are due to changes in single nucleotides, and cataloging of these single nucleotide polymorphisms (also known as SNPs) has become an important endeavor in human genetic studies. These SNPs are also important in determining how genetic variation influences responses to drugs.

Because natural selection requires genetic variation to operate, evolutionary geneticists have a keen interest in the origin and maintenance of variation and have been dealing with patterns of variation in populations since the 1930s. Over the course of the last several decades, these evolutionary geneticists have developed increasingly sophisticated models to explain and predict the fates of genetic variants under a variety of scenarios. Some of these models, particularly those that examine patterns of the correlations of closely linked SNPs, have informed the mapping of genes for complex traits.[17]

In chapters 4 and 5, we'll discuss how these patterns of correlations among genetic variants—what evolutionary geneticists call linkage disequilibrium—can be used to infer the action of various types of natural selection acting on genes. More generally, a major theme of this book will be what we can learn about selection and other evolutionary forces through observations of the patterns of DNA sequences within and among populations and species of organisms.

■ Maladaptations—Evolution Is Not Perfect

Evolutionary biology doesn't just explain adaptation, that is, the goodness of fit of an organism to its environment; it also explains maladaptation. Why do we age? Why does our body wear our as we get older until we are left, as the Bard put it, "sans teeth, sans eyes, sans taste, sans everything"?

As early as the 1950s, evolutionary biologists outlined the solution as to why our bodies break down with age.[18] Consider that even in the absence of loss of function with age, accidents and predators (and pathogens) cause death; by necessity, more individuals must survive to age six years than those who survive to age 60. Because fewer individuals are present at the older age, the power of natural selection declines with the age of the organism.

Now suppose that a mutation arises that causes a greater accumulation of cholesterol. Such a mutation would be harmful in the body of a 60-year-old individual, because it leads to hardening of her arteries. But that same mutation could be selectively favored if it allowed a toddler to grow faster. Even in the absence of a beneficial effect at an early age, slightly deleterious mutations whose effects are only manifest at a later age can accumulate because the action of random genetic drift overpowers the weak selection of such mutations. Geneticists can point to a substantial and growing list of genetic variants that influence the likelihood of obtaining diseases of old age; for instance, variants at the *Apolipoprotein-E* gene affect the propensity to acquire Alzheimer's disease, as well as coronary heart disease.[19]

Our bodies also contain maladaptations that are the remains of the burdens of history. Vestigial organisms—the human appendix is a well-known example—once served a useful function but now no longer do so. They remain ghosts of adaptation past. Obesity in the developed world today is a consequence of our fat storage systems having evolved in an environment in which highly caloric sources of food were less prevalent than they are today.

■ An Inordinate Fondness

The genome also provides much evidence that we are not well designed. Genes that code for proteins make up only about 1.5% of the total genome. If you were to add to these coding genes a generous estimate of the extent of the sequences surrounding these genes that play roles in their regulation, you would be up to no more than 5%. The remaining 95% of the genome is not involved with making proteins; most of this noncoding DNA is made up of highly repetitive sequences and mobile genetic elements that replicate and jump around the genome. Although a small percentage of these elements may actually contribute to the function of the organism, the vast majority of them do not. In fact, the aggregate total of such elements may, like barnacles on a whale, be deleterious. But we can't easily get rid of them.[20]

The mid-twentieth-century polymath and evolutionary biologist J.B.S. Haldane—who as a popularizer of science was the Carl Sagan and Stephen J. Gould of his era—was once asked what we could tell about the Creator from observing patterns of nature. Haldane supposedly replied, "He has an inordinate fondness for beetles."[21] Haldane was referring to the immense diversity of beetles, which make up by some estimates 40% of all known species of insects. Were Haldane around today, he might say that the Creator has an inordinate fondness for *Alu*, which is a particularly abundant mobile genetic element whose combined sequence length makes up 10% of our genome.

■ Evolution Outside of Biology

The principles of evolution are important even outside of biology. Engineers increasingly use genetic algorithms that mimic the processes of evolution to solve difficult design problems. For example, the genetic algorithm approach has been used successfully to design ever smaller and more complex computer chips.

During the late 1980s, Tom Ray, then a young ecologist at the University of Delaware, created an artificial world that he called Tierra, which consisted of a number of computer programs that competed with one another for resources. These computer programs had the capacity to replicate themselves by carrying out a series of instructions that were 80 bytes long. At first, nothing unusual happened as Tierra filled up with copies of the original program. But Ray had been smart, and had allowed for small changes in the code to occur occasionally. These changes, which acted in the same way as random mutation, permitted the evolution of the Tierra system. Although most changes were either innocuous or harmful, some changes, such as elimination of redundant steps, allowed for greater efficiency and thus more offspring. In addition to mutation, Ray also built "death" into the Tierra system, killing off the oldest or most defective program at given intervals.[22]

The combination of faithful reproduction with occasional error (mutation), competition for resources, and variation in success found in Ray's artificial system is the same combination that Darwin saw as the key to evolution via natural selection. And as in the biological world, Ray's computer "organisms" evolved, and did so in ways that he had not imagined. Some programs devised ways to usurp the machinery of other programs to reproduce themselves, and thus became parasites. These parasites, because they no longer required their own reproductive machinery, shrank in size and were therefore able to pump out more descendants in a shorter time. Ray also noticed that the system would go through long periods without much change, punctuated by brief intervals of intense change, an evolutionary pattern similar to those seen in the biological world.

Building on Ray's results, other researchers examined the similarities and differences between Tierran and bacterial evolution. When mutation rates are low and programs are unable to access other programs' files (eliminating the possibility of parasitism evolving), the Tierran programs behave much like bacteria adapting to a new environment. In this pattern, called periodic selection, which is seen in both bacteria and the computer programs, the population is usually dominated by a single variant, with all other variants being at low frequency. Periodically (hence the name), a mutation, which had been derived from the predominant variant, would quickly rise in frequency and would eventually become the new dominant variant.

■ Legal Applications

In recent years analysis of evolutionary relatedness has been used in legal matters. In the case of the *State of Louisiana v. Richard J. Schmidt*, Dr. Schmidt, a gastroenterologist, was accused of attempted murder.[23] Schmidt had been charged with injecting his ex-girlfriend with blood from an HIV-infected patient under the doctor's care. A vial of HIV-infected blood had been found in Schmidt's office, but was that the cause of the victim's HIV infection?

Because HIV mutates and evolves so quickly, inferring whether the victim had indeed received HIV from the suspect required more than just comparing the DNA sequence of the HIV in the victim and that of the vial. Sequences of viruses from a local population of HIV-infected individuals were obtained. With these data, the researchers employed sophisticated methods of inferring evolutionary relatedness. They were able to determine that HIV sequences from the victim clustered with HIV sequences from the vial and not those from a local population of HIV-infected people. Moreover, the researchers inferred that the victim's HIV was derived from that of the vial (and not the other way around). This evidence helped convict Schmidt of attempted murder in the second degree.

While I was writing this book, I served on a jury trial in a criminal case. The actual trial was much like one sees in television shows such as *Law and Order* and in John Grisham's books, except that real-life court cases are more repetitive. My experience led me to think about similarities and differences between scientific and legal reasoning. In science and in law, decisions are based upon evidence. Evidence in both systems can be obtained from direct observation or from circumstantial evidence (reasonable inferences from observations). For example, no one actually saw Timothy McVeigh at the Murrah Federal Building in 1995, but he was convicted and eventually executed because of strong circumstantial evidence tying him to the crime. In most fields of science as well, strong circumstantial evidence is at the base of a large proportion of scientific conclusions.

In law, having multiple lines of evidence back up the same conclusion can increase the strengths of one's claims. In the McVeigh case, traces of explosives residue were found on his clothing when he was arrested 75 minutes after the bombing. His fingerprints were found on a receipt for a large quantity of ammonium nitrate, one of the ingredients used to make the bomb. McVeigh also had a false diver's license made out to Bob Kling, the name used to rent the truck that carried the bomb. As a result of these and other pieces of evidence, prosecutors charged McVeigh with murder and the jury found him guilty.[24]

In science, too, having multiple lines of evidence strengthens a case. Common evolutionary decent with modification via Darwinian processes is supported by a tremendous body of evidence coming from all areas of biology. Darwin based his claims from knowledge of geology and the emerging fossil record, breeding, comparative development, comparative embryology, and patterns of distribution of plants and animals across space and time, among other fields. New information from these fields over the last 150 years continues to support the core of Darwin's theory. In addition, scientists from areas of science that Darwin could not have dreamed of—for instance, comparative developmental genetics, comparative biochemical studies, and now comparative genomics—all bolster what Darwin had written. To a lesser extent, the same is true about particular claims regarding the evolution of a particular species or trait. Strong cases are often built on multiple, mutually reinforcing lines of evidence.

■ Coda

We live in an age of biology. Hardly a day goes by without news of a new discovery in biology or medicine—the sequencing of a genome of yet another organism, the production of a new drug, the finding that variants at particular gene affect the likelihood of acquiring a disease. Today, we face new threats and the reemergence of old threats, including the looming specter of bioterrorism as well as outbreaks of disease. Responses to these threats and other challenges require a strong scientific infrastructure and science education. Evolution, both as the central organizing principle of the life sciences and as the foundation for numerous practical uses of science, must be a primary focus in the science curriculum at all levels.

Evolution should be a larger part of the curriculum at the college and high school level for reasons aside from the practical ones outlined above. Darwin's evolution by natural selection is one of few scientific theories that fundamentally changed the way we view the world. Moreover, study of Darwinian evolution is an excellent way to teach science as a way of knowing.[25] As noted above, much of science—including but certainly not limited to evolutionary biology—rests upon the proper use of making inferences from observations. Exposing students to the work of Darwin and his intellectual descendants helps show students how science works.

Why Intelligent Design Is Not Science

Intelligent design may be interesting as theology, but as science it is a fraud. It is a self-enclosed, tautological "theory" whose only holding is that when there are gaps in some area of scientific knowledge—in this case, evolution—they are to be filled by God.

—Charles Krauthammer (Krauthammer, 2005)

■ What's the Matter with Kansas?

Although the struggle over how evolution is taught in public schools is being waged in several battlegrounds across the United States, the state of Kansas has received the most attention. In 1999, the Kansas State Board of Education voted to eliminate reference to evolution and an ancient Earth from its standards. After considerable outcry and widespread mockery of Kansas, several of the anti-evolution members of the board were voted out of office the following year. But the political climate changed again; in subsequent years, the Kansas electorate voted in a majority of anti-evolutionists onto the board. Although the board of education does not mandate curriculum, the standards that it adopts are used in the creation of statewide assessment tests and thus greatly influence what is taught in the classroom.

On November 8, 2005, the board voted 6 to 4 to approve a proposal regarding the state's standards for science.[1] Although these new standards state that students should learn evolution, they also encourage students to keep an open mind about evolution. The standards assert that Darwin's theory of evolution contains gaps and that recent evidence challenges the theory's validity. What's wrong with that? Students certainly should be taught critical thinking skills, including the ability to weigh evidence while keeping an open mind when engaged in all matters of inquiry. The problem with the board's standards is that it singles out only evolution for such treatment. The standards don't mention weighing the evidence and keeping an open mind

17

about Newton's laws of motion, the theory of gravity, or the properties of chemical bonds. The reasonable inference that students and others would draw based on these revised standards is that evolution as a science is on shaky grounds. And yet, evolution is as solid as any other branch of science.

Most disconcerting to the scientific community, the board rewrote the definition of science such that it is no longer restricted to searching for natural explanations for observed phenomena. These standards allow "science" to seek supernatural causes. Here is a case of a legislative body, with no scientific expertise and no training in the practice of science, altering the very definition of science to suit its religious and political agendas. This body has done so in clear disregard for the expressed opinion of the National Academy of Sciences, the National Center for Science Education, and several other scientific and educational agencies. Furthermore, this change in the definition of science goes against the principles of the Enlightenment and several centuries of scientific thought.

Is the board advocating intelligent design, the proposition that an "Intelligent Designer" must be the cause of some or all aspects of living organisms because these features are too complex to have arisen by natural processes? Although the standards' authors took care to claim that they were not endorsing intelligent design, the language in the standards regarding evolution mirrors that of the Discovery Institute, the think-tank that has been the major player in promoting intelligent design. We will look further into the Discovery Institute and its agenda later in this chapter.

Not all Kansas politicians agreed with the board's decision. Four of the ten board members opposed the change, as well as the governor of Kansas, Kathleen Sebelius, who observed that the board's decision to weaken the standards would hinder the economic competitiveness of her state.[2]

■ Intelligent Design on Trial

Kansas is far from the only location wherein battles about teaching evolution and intelligent design have been waged; indeed, the intelligent design movement and the "evolution wars" are national, if not international, phenomena. One hotspot during 2004 and 2005 was Dover, Pennsylvania. Situated in southeastern part of the state, the town of Dover became the unlikely site for the most publicized trial revolving around teaching of evolution since the Scopes decision.

Events began in fall 2004 when the Dover school board ordered ninth-grade biology teachers to read a statement presenting intelligent design as an alternative to Darwinian evolution. In a letter to the board, the teachers announced that they could not in good conscience read that declaration because doing so would "knowingly and intentionally misrepresent subject matter or curriculum."[3] The school board eventually backed down, letting the teachers off the hook. In place of the teachers, Dover School Superintendent

Richard Nilsen and his assistant read the statement to all nine ninth grade biology classrooms.

Among other things, this disclaimer stated that Darwin's theory is not fact, and "Gaps in the Theory exist for which there is no evidence." The disclaimer also noted that "Intelligent Design is an explanation of the origin of life that differs from Darwin's view. The reference book, *Of Pandas and People*, is available for students who might be interested in gaining an understanding of what Intelligent Design actually involves."[4]

Several of the students' parents objected; led by Tammy Kitzmiller, they sued the school board, setting the case in motion. Judge John E. Jones, a conservative Republican appointed by President George W. Bush, presided over this trial, formally known as *Kitzmiller et al. v. Dover Area School District et al.*, which took place in fall 2005. Since the 1980s, previous courts had ruled that the teaching of "scientific creationism" in science classes in public school was unconstitutional because such teaching advanced the establishment of religion in the public square, forbidden under the Establishment Clause of the First Amendment. At issue in Kitzmiller was whether the Dover school board's attempt to bring intelligent design (ID) into the public classroom also violated the Establishment Clause.

We will return to both Dover and Kansas later in this chapter, but first let's investigate exactly what the proponents of intelligent design claim. More important, what is the basis for their positions? By addressing these questions we can determine whether ID is science, a form of creationism, or something else.

■ Intelligent Design and Behe's Black Box

Although University of California at Berkeley law professor emeritus Philip Johnson has been popularizing the term "intelligent design" and the ideas behind it since 1991, the most important figure in the ID movement now is arguably Michael Behe. As professor of biochemistry at Lehigh University (ironically, not far from Dover, Pennsylvania), Behe, unlike nearly all of the other proponents of ID or previous forms of "creation science," is a real professional biologist. Not only did Behe earn a doctoral degree in biological sciences, but he has also led an active research career, publishing in peer-reviewed scientific journals and attempting to obtain research grants. Thus, from the perspective of the general public, he appears to speak from authority. Behe was also the star witness for the ID side in the *Kitzmiller v. Dover* case.

Behe's first major foray into the ID debate was his 1996 book, *Darwin's Black Box*.[5] This book differed from previous ID works in that Behe's criticisms of orthodox evolutionary biology and his argument for intelligent design were based on examining biochemical properties of organisms. Given that *Darwin's Black Box* is the *only* ID book written by a professional biologist and that its claims are presented as being based on biochemistry, the focus of the scientific aspect of this chapter will be on responding to Behe's claims.

So, what does Behe say? First, he accepts an ancient Earth. Indeed, in *Darwin's Black Box*, he states: "For the record, I have no reason to doubt that the universe is the billions of years old that physicists say it is."[6]

Although this statement is certainly a break from those who believe the Earth is only a few thousand years old, these young-earth creationists are not the only type of creationist. It is important to realize that several variations of creationist thought coexisted throughout the twentieth century; not all of these hold the view that the Earth was created less than 10,000 years ago. For instance, popular depictions of the 1925 Scopes "monkey" trial, as epitomized in the play (and movie) *Inherit the Wind*, by Jerome Lawrence and Robert E. Lee, present Clarence Darrow's cross-examination of William Jennings Bryan (the progressive presidential candidate turned anti-evolution crusader) as the turning point of the trial. According to the play, Darrow got Bryan to admit that each of the days in Genesis could represent periods that span millions of years, and thus exposed the intellectual vacuum of the creationists. In reality, however, this was not an embarrassing admission for Bryan. Like many, and perhaps the majority, of anti-evolutionists at the time, Bryan subscribed to what is known as the "Day-Age" creationist view, wherein the days of Genesis represented ages of perhaps immense time.[7] Not until after World War II was the leadership of the creationist movement mainly made up of advocates of "Young Earth" creationism.

Behe goes on to say, "Further, I find the idea of common descent (that all organisms share a common ancestor) fairly convincing, and have no particular reason to doubt it."[8] Behe's acceptance of common descent, even if it is lukewarm, is indeed a departure from much of creationist thought.

The quarrel Behe has with Darwinian evolution is with the mechanism. He says, "Although Darwin's mechanism—natural selection working on variation—might explain many things,...I do not believe it explains molecular life."

Behe contends that molecular life—the biochemical machinery of the cell—is too complex to originate from natural selection and other evolutionary forces operating on existing variation (that is, replenished by new mutations and recombination). He is not concerned with complexity in general but with a specific type of complexity that he calls "irreducible complexity." Due to the importance Behe places on "irreducible complexity," it is instructive to get at the meaning of his term. To Behe, a system is irreducibly complex if it is "composed of several well-matched, interacting parts that contribute to the basic function, wherein the removal of any one of the parts causes the system to effectively stop functioning."[9]

Behe uses the analogy of a mousetrap to explain irreducible complexity. In a standard mousetrap, all of the parts are required in order for the trap to work (catch mice). Removal of any essential part would result in an inoperable mousetrap. Thus, a mousetrap could not have evolved from the successive accumulation of components; it must have been designed. Likewise, Behe argues, various biochemical systems in organisms are irreducibly complex. Removal of any one of the components results in nonfunctionality; thus, the biochemical system could not have evolved from the successive accumulation of its components. Therefore, those irreducibly complex biochemical systems must have been designed.

Behe presents several examples of what he claims to be such irreducibly complex biochemical systems; most notable is the blood-clotting system of vertebrates, about which Behe devotes more than 20 pages of his book. Behe is correct; the clotting cascade and many other biochemical systems are remarkably complex. Behe is also a gifted writer; the marvelous detail that he vividly and cogently illustrates forces the reader to come to grips with the intricacy of biochemical assemblages.

But Behe's thesis is incorrect. The remarkable complexity of biochemical systems can and does evolve via Darwinian processes. Before we delve into the elaborate biochemical details, let's explore antecedents of Behe's argument and how biologists have dealt with them.

■ Breaking Through Irreducible Complexity—The Eyes Have It

First, the form of Behe's argument is not new.[10] Two hundred years before Behe, the theologian William Paley argued that just as watches show signs of being designed, so do organisms. Darwin was very much aware of Paley and this "argument from design," and indeed had been impressed by Paley as a young theology student. As an older man, however, Darwin amassed evidence and used reason to show that this argument was not necessary: natural selection along with other natural processes could lead to organisms that appear to be designed.

After Darwin published *The Origin of Species*, his critics charged that complex systems could not evolve via his mechanism of natural selection. Over the past century and a half, skeptics of Darwin have often pointed to the vertebrate eye as an example of a system too complex to have evolved by Darwinian processes. A mid-nineteenth-century author put it best: "To suppose that the eye, with all its inimitable contrivances for adjusting the focus to different distances, for admitting different amounts of light, and for the correction of spherical and chromatic aberration, could have been formed by natural selection, seems...absurd in the highest possible degree."[11] Or, stripping away the flowery Victorian prose, what possible use is half an eye? The answer is plenty. As we shall see, in the kingdom of the blind, the animal with even a slightly functioning eye is king.

I must confess that I've just pulled a small rhetorical trick. The passage I've quoted above is not from one of Darwin's critics, but from Darwin himself. Darwin struggled with the problem of complex organisms such as the eye, but he had formulated a solution (that creationist authors seldom provide). Let's examine more of what Darwin said about eyes:

> Reason tells me, that if numerous gradations from an imperfect and simple eye to one perfect and complete, each grade being useful to its possessor, can be shown to exist; as is certainly the case; if further, the eye slightly varies, and the variations be inherited, as is likewise certainly the case; and if such variations should ever be useful to any animal under changing conditions of life, then the difficulty of believing that a perfect and complex eye could be formed by natural selection, though insuperable by our imagination, cannot be considered real.

In other words, the objection fails if one can show that the eye could have arisen via small, successive steps that are each potentially beneficial. Although it would be wonderful to know exactly how the eye formed, all we need to show is a plausible route by which evolution could have taken. This will serve as a starting place for further inquiry.

Image-forming eyes evolved independently literally dozens of times throughout the animal kingdom. As we shall see, the types of eyes that different animals use vary extensively. Our eyes and those of many vertebrates contain a flexible lens that focuses light onto the retina, a thin film that contains light-sensitive cells. In contrast, insects use compound eyes; such eyes contain numerous simple facets, each with photosensitive cells and a miniature lens. Although the compound eyes result in a pixilated image, these eyes are highly motion sensitive.[12]

As we consider how these eyes could have evolved, let us, as Maria from *The Sound of Music* suggested, start at the very beginning. In the beginning (as far as eyes are concerned) was the photocell, a specialized cell containing at least one pigment that is able to collect tiny packages of light known as photons. Even some single-celled organisms, like the freshwater protozoon *Euglena*, have regions of their cell that are sensitive to light.

Photocells themselves vary greatly in complexity and in their ability to capture photons. All things being equal, a more sensitive photocell will be favored over a less sensitive one. Thus, it is easy to see how increased complexity and thus increased functionality of photocells can be built up through small, successive steps that Darwin proposed.

Although photocells are used in all of these vastly different types of eyes, these photocells differ in details. For instance, the layers used to collect photons in squid photocells are arranged like old 45-rpm records around a hollow tube. In contrast, in the photocells of mammals, these layers are arranged like a stack of compact disks. Such differences in detail reflect the fact that evolution often "solves" the same problem in disparate organisms by slightly different routes. Note that "solves" is in scare quotation marks; evolution does not consciously solve problems, but often the results of the unconscious process of evolution appear to be solutions to problems.

How do we get from a photocell to an eye? Through small, successive steps. Having more than a single photocell is an improvement to having just one. Multiple photocells could be produced from small changes in development. Increasing the number of photocells—through small, successive steps—could lead to the formation of a thin layer of photocells. We can think of further improvements, all of which could arise from small steps. Being able to detect the direction from which light is coming is an improvement over just the ability to see whether there is light. Such an improvement could arise from folding of the photocell layer. Diverse animals—the clam, the flatworm, and the limpet—have cup eyes that are further modifications of such folded layers of photocells. Richard Dawkins discusses further evolution of the eye in more detail in his excellent book about how complex structures can and do evolve, *Climbing Mount Improbable.*

In a mathematical model, Dan Nilsson and Suzanne Pelger[13] demonstrate that via natural selection, a flat layer of photocells can evolve into something that looks like a fish eye. This process occurs through a few hundred to a few thousand steps, but it can be completed within less than half a million generations. Their model makes various assumptions about mutation rates, strengths of selection, and the genetic basis of the traits, but all of these appear reasonable. Half a million generations is not all that long in evolutionary time. Given a few generations per year, this process can be completed in about 100,000 years, which is but a flicker of an eye in geological time.

Features of the diverse eyes that actually evolved provide evidence that these eyes have evolved in a historical context, rather than having been designed. Located in our retinas is a space that lacks photocells. This is where the optic nerve, the cable that conducts signals from the retina to the brain, connects. The result is a "blind spot." The reason that we don't usually perceive this blind spot is that we have two eyes and our brains are equipped to fill in the missing gaps. This way of hooking up the optic nerve is a historical contingency; a better eye could have been designed that had no blind spot. Indeed, such eyes have evolved; in the octopus eye, for example, the optic nerve attaches to the underside of the retina, and there is no blind spot.

■ Biochemistry—Step by Step

We've shown that the eye can be build up from humble beginnings through small successive steps. The same can be said for just about any other complex organ that has been intensively studied. We may not know all the intricate details of their evolution, but evolutionary biologists can show that they could reasonably have evolved through small, successive steps. So, what about Behe's realm; is biochemistry any different?

The short answer is no. Whereas modern biochemistry did not begin until the twentieth century, the study of morphology extends back to the early nineteenth century, if not earlier. Because detailed morphological studies were feasible long before similar biochemistry studies, in general we know more about morphology than we do biochemistry. In cases in which extensive studies have been performed, however, evolutionary biochemists can show that these biochemical systems can also be built by small, successive steps.

Let's start with Behe's favorite system: the cascade of reactions involved in blood clotting.[14] Circulating in relatively high concentrations in my blood, your blood, and the blood of all vertebrates is a protein, called fibrinogen. It has the potential to form clots because it is predisposed to stick to copies of itself. The formation of clots in blood vessels is generally bad, but fortunately, a small chain of amino acids called fibrinopeptides on the exterior of fibrinogen prevents fibrinogen from sticking to itself.

So how do clots form? At the site of a wound, clotting starts when an enzyme called thrombin cleaves off the fibrinopeptides from fibrinogen. Once the fibrinopeptides are removed, these proteins, now called fibrin, stick to one another. Thrombin freely circulating through the blood would lead to the conversion of fibrinogen into fibrin, and thus clots in blood; so thrombin itself has to be activated. Indeed, thrombin naturally exists as prothrombin, which can be activated by another enzyme known as factor X at the appropriate time. The X in factor X doesn't signify unknown, but rather the Roman numeral for ten.

Factor X is of particular interest because it can be activated in two ways. The presence of tissue factor—found in most tissues but not normally blood— is released when a blood vessel breaks within a tissue. Exposure of blood to air leads to cell surface damage. Both the cell surface damage and the tissue factor, working through intermediaries, lead to the activation of factor X.

Why is this process so baroque? The complexity of the system arises because the various steps are used to amplify the response in the cascade. By adding steps to the cascade, a stronger response to a wound can be made. It's similar to the old slogan in a shampoo commercial some years ago where one person who liked the shampoo told two friends, and those friends told two friends, and so on, and so on. So, the seemingly baroque cascade appears to be advantageous.

Now, how could this elaborate pathway evolve? A major clue is that many of the clotting enzymes—including prothrombin and factor X—are quite similar to one another. This similarity is a strong indicator that these genes arose via a series of gene duplications. Other patterns present in the sequences of the genes that produce these enzymes lend further support to the possibility that these genes arose from duplications. In addition, thrombin shares similarity in sequence with an enzyme called trypsin, which is involved in the cleavage of proteins in the pancreas. The pancreas and the liver (which produces the enzymes involved in blood clotting) are "sister organs," that is, the cells that make up the liver and those that make up the pancreas are derived from the same cells in the early development of the organism.

These data all support a step-wise mode of evolution of the blood clotting system. In outline, the system began as a simple one (fibrinogen and thrombin are among the earliest components of the system) and gradually became more complex as gene duplications allowed for new components of the system. The addition of a new component gave an advantage to the organism by providing amplification of the signal, whereby a small change in the component early in the clotting system could allow for much larger changes later in the system. Such an addition probably came at a cost because the functions of the new gene were not fully in tune with the rest of the system. This cost, however, was smaller than the added benefit that the new gene provided. Subsequent changes that integrated the new gene's functions with the system as a whole could then minimize the cost, thereby providing a selective advantage.

One may predict from such a model that some components of the system would have appeared later than did others. By comparing the components present in different organisms, biologists can test these predictions. To give a specific example, three components (prekallikren, factor XI, and factor XII) are thought to be latecomers to the cascade and thus would not be found in distant relatives of mammals and other terrestrial vertebrates. When the pufferfish genome was sequenced, Russ Doolittle and his colleagues looked at which clotting components it contained.[15] Although the pufferfish had most

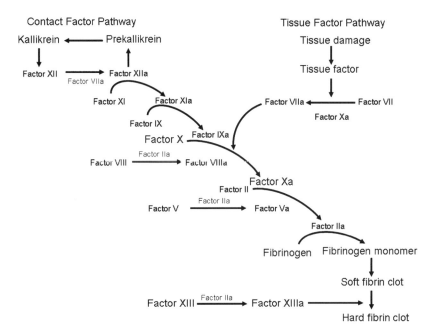

Figure 2.1
The blood clotting cascade. The clotting reaction can begin at either the contact pathway or the tissue factor pathway, and factor X is involved in each step. The "a" denotes the activated form of a factor (e.g., Factor Xa is the activated form of Factor X).

of the components of the cascade, it lacked prekallikren, factor XI, and factor XII—as expected from the distributions of components in related organisms. Intelligent design can't even make such predictions.

Tellingly, Behe, in a 2002 dialogue with Kenneth Miller, admitted that the blood clotting system is not irreducibly complex like a mousetrap.[16] One reason it isn't irreducibly complex is that some animals (dolphins are a key example) exist that are missing parts of the system. Redundancies are built in.

The blood clotting system is not the only complex biochemical system for which a plausible step-by-step scenario for its evolution has been worked out. In the fruitfly, *Drosophila melanogaster*, females have two X chromosomes and males have one X chromosome. Sounds simple, doesn't it? The actual pathway that determines sex in this fly, however, contains several steps and is another example of a remarkably baroque biochemical system. Among other features, the pathway has several alternative splicing patterns, wherein one RNA can be spliced in different ways; the end result is that many different proteins can be manufactured from the same gene. Genes in this pathway also have multiple sites at which they can be regulated by other genes. In addition, the proteins produced by some genes regulate their own production. This scenario definitely seems complex, and it is. It's also likely that the removal of any one of several parts of the present pathway would lead to the production of flies that are neither male nor female, and are thus evolutionary dead-ends. Nonetheless, a plausible mechanism exists that can account for this pathway's evolution.

Comparative evidence from looking at the patterns seen in a variety of other insects strongly suggests that the complex pathway in *Drosophila* arose in a series of steps from a much simpler "one-switch" system. In 2004, Andrew Pomiankowski and his colleagues proposed a model that attempts to explain the evolution of the major features of this pathway in terms of population genetics. The details of this model are the beyond the scope of this book, but the formation of such a model generates predictions that can be tested.[17] Such models also can point to new directions of biological study.

■ God of the Gaps

Whether for the clotting cascade, sex determination pathways, or any one of a number of other complex biochemical systems, evolutionary biologists are able to devise plausible models for how such systems evolved. These models are subject to testing by a variety of means—whether through laboratory experiments or through the observation of related individuals. It is true that gaps remain in our understanding about numerous aspects of biology, including biochemistry. But that doesn't mean that those gaps will not be explained through the course of further inquiry. The key point is that these models lead to further scientific inquiry, in contrast to the consequences of throwing one's arms up and pointing to an intelligent designer to explain the gaps. Just because we don't know everything doesn't mean that we don't understand the principles involved.

In one of my favorite scientific cartoons by Sidney Harris, two scientists are looking at a blackboard filled with mathematical symbols and other "scientific looking" writing. In the middle of the blackboard the phrase "then a miracle occurs" appears. One of the scientists says to the other, "I think you should be more explicit here in step two."[18]

Invoking the phrase, "Then a miracle occurs," to processes that we do not understand is not a scientific statement. This is an example of a "God of the Gaps" argument, wherein one argues for a supernatural explanation based on gaps in our scientific understanding. And this is precisely what the proponents of intelligent design are doing—despite their denials—when they invoke "Design." One obvious predicament with the "God of the Gaps" argument is that gaps, as we have seen, can be filled in and explained by scientists. The larger problem, however, is that such an argument is counter to scientific reasoning; reliance on a faith-based explanation does not advance fact-based understanding.

■ Facts and Theory

During the 1980 presidential campaign, former President Ronald Reagan was asked about evolution. He replied, "Well, it's a theory, it is a scientific theory only, and it has in recent years been challenged in the world of science and is not yet believed in the scientific community to be as infallible as it once was believed."[19] Over the next two decades, subsequent critics of evolution have used similar language in attempts to dismiss evolution. After all, the popular perception of the word "theory" is "educated guess." Or as the biochemist and science-fiction writer Isaac Asimov said in the same year as Reagan's quote, "[Creationists] make it sound as though a 'theory' is something you dreamt up after being drunk all night."[20]

In formal, scientific use, a theory is much more than an educated guess. It is also more than a hypothesis; it is a coherent body of well-tested hypotheses. A theory is a set of statements that (1) are interconnected, and internally consistent; (2) are based on observable and circumstantial evidence ("the facts"), (3) explain a wide variety of observations, and (4) have been tested by reference to the observable facts.

The theory of evolution is upheld by, and explains, numerous well-supported pieces of evidence that we accept as fact. It is a fact that humans and chimpanzees are closely related species that diverged from a common ancestor between five and seven million years ago. It is a fact that humans, chimpanzees, mice, and kangaroos belong to the class of vertebrates known as mammals, and that these organisms are placed in that class based on sharing particular characteristics (for instance, are warm-blooded, possess mammary glands, and have hair). It is a fact that mammals that fly are found on oceanic islands, whereas nonflying mammals generally are not. These facts both support and are explained by common descent with modification via natural selection and other natural forces—in other words, Darwinian evolution.

Evolutionary theory is not just a theory; as the central organizing principle of the life sciences, it is one of the most encompassing theories in all of science. Many smaller theories are contained within the theory of evolution. Existing evidence supports these smaller theories by varying degrees. Researchers are continually testing theories by using a variety of methods— observations, manipulative experiments, mathematical theory, computer simulations. In this testing, some of the smaller theories gain support and others lose support. By this process, the theory of evolution as a whole is refined and adjusted based on new evidence. We will see some examples of refinements and even re-refinements of evolutionary theory in the chapters to come.

Although evolution is both facts and theory, intelligent design is neither. It is incorrect to use the phrase (except with scare quotation marks) "intelligent design theory." Reference to an "Intelligent Designer" is not an explanation. An Intelligent Designer cannot be tested. One can speculate about the nature of such a Designer—whether it acted once and has been silent since or whether it continually acts (or somewhere in between)—but such speculation is not testing; it is not science.

■ ID, DI, and the "Wedge" Agenda

Despite the claims by the ID advocates that they are doing science, it is evident that they do not practice accepted modes of scientific reasoning and empirical validation. It is also clear that these ID proponents are guided by and are propagating faith-based religious and political agendas. The best evidence that they have these agendas comes from the so-called "Wedge Document" put out by Seattle's Discovery Institute.

Formed by Bruce Chapman in 1990, the Discovery Institute (DI) was originally a group of mainly business-oriented, socially liberal Republicans. Focusing on telecommunications and issues local to the Seattle area, the early DI was described as having "a forward-looking, futuristic, and intellectually contrarian" vibe.[21] The DI's initial purpose is reflected in its current mission statement: "The Institute discovers and promotes ideas in the common sense tradition of representative government, the free market and individual liberty."[22]

During the 1990s, however, DI and Chapman (who still serves as its president) moved steadily rightward politically and religiously. DI, through its Center for Science and Culture (CSC), now serves a major source of funding and other support for ID proponents and other critics of evolution. The Discovery Institute's CSC has become the headquarters of a movement that, as described by Barbara Forrest and Paul Gross, "seeks nothing less than to overthrow the system of rules and procedures of modern science and intellectual footings of our culture laid down in the Enlightenment."[23]

On what basis did Forrest and Gross draw these conclusions? From the DI's own "Wedge Document," which surfaced on the Internet in early 1999. (DI admitted to owning the document, but only in 2002). This document states, "Discovery Institute's Center for the Renewal of Science and Culture seeks nothing less than the overthrow of materialism and its cultural legacies."[24] It views materialism—the tenet that all that exists is material or physical—as the root of all evil, a position taken by many creationists of the nineteenth and twentieth centuries. Materialism is closely aligned to, but not the same thing as, naturalism, the search for only natural explanations in explaining phenomena. As we shall see, naturalism exists in more than one form.

The "thin edge" of this wedge was the publication of scholarly works on intelligent design by Michael Behe, William Dembski, Philip Johnson, and a few other members of DI's board. The next part of their strategy would then be attempts to sow doubt about Darwinian evolution among an unsuspecting public (especially school boards), presenting ID an alternative. As Forrest and Gross chronicle, the Discovery Institute, through its CSC, has been actively working with school boards in several states (notably Kansas, Ohio, and Washington State). In the case of Kansas, they succeeded in rewriting the standards to place undue doubt regarding what we actually do know about evolution.

The Wedge authors also collaborated with conservative religious leaders such as James Dobson of Focus on the Family. In addition, they conspired with politicians, particularly in the "right wing" of the national Republican Party, to advance their agenda. In particular, Senator Rick Santorum of Pennsylvania almost succeeded in writing Wedge-document language regarding evolution into the "No Child Left Behind" act of 2001. Specifically, Santorum's amendment was lifted from the legal playbook of the ID movement, "Teaching the Controversy: Darwinism, Design, and the Public School Curriculum," written by several DI members. Although the Santorum amendment was ultimately defeated, the nonbinding conference report associated with the bill used similar but diluted language.[25]

The agenda of those who champion ID is not just a religious one but one that puts forth a particular religion. Although in the attempt to gain credibility, they will sometimes claim that the Intelligent Designer could be an alien, they really aren't fooling anyone. Their Designer is clearly the Christian God. Their very words betray those motives. For instance, here are a few quotes from William Dembski, a mathematician and philosopher at Baylor University in Texas and one of the leading ID proponents:

My thesis is that all disciplines find their completion in Christ and cannot be properly understood apart from Christ....[26]

The point to understand here is that Christ is never an addendum to a scientific theory but always a completion....[27]

Christ is indispensable to any scientific theory, even if its practitioners don't have a clue about him.[28]

■ Creationism and Conservatism Need Not Be Linked

Although the Discovery Institute has collaborative ties with the right wing of the Republican Party in the United States as well as the "Religious Right," it is not the case that creationism (in either its current or previous incarnations) and conservatism must be linked.

Several prominent conservative intellectuals have come out as defenders of evolution as a science and as strong critics of the ID movement. Charles Krauthammer, who was quoted at the start of the chapter, is a long-time conservative columnist for the *Washington Post*. He has called the current evolution battle "so anachronistic and retrograde as to be a national embarrassment."[29] Krauthammer is not the only conservative pundit to have vociferously denounced the claim that ID is science. George Will denounced the Kansas school board's redefinition of science and its proclamation that evolution "is not a fact."[30] "But it is," Will responded. After quoting Thomas Jefferson, who said, "It does me no injury for my neighbor to say there are twenty gods, or no God. It neither picks my pocket nor breaks my leg," Will notes, "But it is injurious, and unneighborly, when zealots try to compel public education to infuse theism into scientific education."

It is also worth noting that, at least in the past, there have been prominent liberal creationists. William Jennings Bryan, a Populist, was considerably to the left of every major candidate for president in the last half of the twentieth century. He was an outspoken advocate for improving the conditions of low-wage workers, the suffrage of women, and the reform and regulation of big business. A committed pacifist, Bryan resigned as President Wilson's secretary of state as the country inched toward entry into World War I. The brutality of the war and the strife at home solidified Bryan as a crusader against evolution. To some extent, Bryan's real beef was not with Darwinism, but with social Darwinism, the misapplication of Darwinian principles to justify racism, social inequality, and imperialism—all anathema to Bryan's progressive and Christian views.[31]

■ Religion and Science Coexisting

A theory is a well-established concept that is confirmed by further scientific discoveries and is able to predict new discoveries. The Big Bang theory and cosmic evolution are confirmed by discoveries in physics ranging from the smallest known particles of matter to the processes by which galaxies are formed. Biological evolution is a web of theories strongly supported by observations and experiments.

The paragraph above, though an accurate and well-written description of scientific theories and of biological evolution as a scientific theory, did not come from a scientist. Instead, it is part of the official position of the

Episcopal Church on science and religion.[32] Obviously, they think one can accept evolution and still be a good Episcopalian.

Pope John Paul II embraced evolution as being more than "mere hypothesis" in 1996. After John Paul's death and the election of Pope Benedict XVI as his successor, many have wondered where the Catholic Church stands on evolution. In July 2005, shortly after Benedict's installation, Cardinal Christoph Schönborn wrote an op-ed to *The New York Times* with language that appeared to be ID friendly. For instance, the cardinal wrote:

> Now at the beginning of the twenty-first century, faced with scientific claims like neo-Darwinism and the multiverse hypothesis in cosmology invented to avoid the overwhelming evidence for purpose and design found in modern science, the Catholic Church will again defend human reason by proclaiming that the immanent design evident in nature is real. Scientific theories that try to explain away the appearance of design as the result of "chance and necessity" are not scientific at all, but, as John Paul put it, an abdication of human intelligence.[33]

But we need not guess blindly as to where Pope Benedict stands on evolution. As a cardinal, Benedict embraced biological evolution. After discussion of the Big Bang, he wrote the following:

> While there is little consensus among scientists about how the origin of this first microscopic life is to be explained, there is general agreement among them that the first organism dwelt on this planet about 3.5–4 billion years ago. Since it has been demonstrated that all living organisms on earth are genetically related, it is virtually certain that all living organisms have descended from this first organism. Converging evidence from many studies in the physical and biological sciences furnishes mounting support for some theory of evolution to account for the development and diversification of life on earth, while controversy continues over the pace and mechanisms of evolution.[34]

In subsequent essays, Cardinal Schönborn explained his view further. In essence, the cardinal stated that he was not objecting to the scientific claims made by Darwinian evolutionary biologists, but rather the philosophical ones.[35] The statements "there are no Gods," "there is no free will," and "matter is all that exists" are all philosophical claims. We will return to this distinction later in the chapter.

The Catholic and Episcopal Churches are not alone in their acceptance of evolution as a science. Over 10,000 members of the clergy of numerous denominations have signed "An Open Letter Concerning Religion and Science" that also states that religion and science are not incompatible. In the letter, they highlight a difference between different types of truths.

> While virtually all Christians take the Bible seriously and hold it to be authoritative in matters of faith and practice, the overwhelming

majority do not read the Bible literally, as they would a science-textbook. Many of the beloved stories found in the Bible—the Creation, Adam and Eve, Noah and the ark—convey timeless truths about God, human beings, and the proper relationship between Creator and creation expressed in the only form capable of transmitting these truths from generation to generation. Religious truth is of a different order from scientific truth. Its purpose is not to convey scientific information but to transform hearts.[36]

Writing in the fourth century, Thomas Aquinas viewed God as the Cause behind natural causes. Early in the twentieth century, the Jesuit-trained priest, Pierre Teilhard de Chardin, espoused an Aquinas-like view of God's hand being behind evolution. A practicing geologist and paleontologist, Chardin was a leading, but far from the only, proponent of theistic evolution. A rich tradition of theistic evolution continues today; in fact, Ken Miller, the star witness for the evolution side in the Dover case, is a theistic evolutionist. Many theologians are also advocates of theistic evolution; some, such as Ted Peters at the Pacific Lutheran Theological Seminary,[37] consider ID to be bad science and bad theology, and want only the best science (Darwinian evolution) taught in all schools (public and religious). Three broad categories of evolutionists exist: the theistic evolutionists, atheistic evolutionists (those scientists who deny the existence of the supernatural), and nontheistic evolutionists (those who make no claim about the supernatural). The differences among these groups are philosophical and theological, not scientific.

■ Naturalism, Science, and Faith

To understand why the differences among atheistic, nontheistic, and theistic evolutionists are not scientific distinctions, we must distinguish between two forms of naturalism. Just about all scientists practice methodological naturalism; that is, they limit themselves to exploring only natural explanations and causes for natural phenomena. This does not mean that they necessarily deny the existence of the supernatural. Some scientists do reject the existence of the supernatural; these scientists are following philosophical naturalism. The difference between following methodological naturalism and advocating philosophical naturalism is one of philosophy, not science. Although scientific methodology requires adhering to methodological naturalism, it does not require philosophical naturalism.

Scientists' suspicions of the supernatural don't necessarily arise because they have an inherent bias toward philosophical naturalism. Rather, they distrust supernatural claims and hypotheses because these have an abysmally poor track record in explaining nature.[38]

Faith is an intensely personal experience, whereas science is universal; the rules and practices of science can be followed by anyone. Indeed, confirmation by independent validation lies at the heart of science. Although science strives for objectivity, faith is subjective by nature. Unlike a scientific dispute, a religious argument cannot be decided on the basis of evidence.

Although many biologists are atheist or agnostic, many are not. At least two of the most prominent biologists who gave us the modern synthesis that united Mendelian genetics and Darwinian evolution in the middle decades of the twentieth century were Christians—and not just go-to-church-occasionally Christians but ones active in their respective Churches. Sir Ronald Fisher, the British biologist who not only formulated much of population genetics but also is considered one of the fathers of biostatistics, was "a devout Anglican who practiced sermons and published articles in church magazines."[39] Theodosius Dobzhansky, best known for testing the ideas of the population genetic theoreticians in both the laboratory and in the wild, was also a believer. He not only belonged to the Russian Orthodox Church but also wrote *The Biology of Ultimate Concern* about the intersection of science and religion. In a review of a creationist book of the 1940s, Dobzhansky wrote, "What a pity that some otherwise efficient intellects are endowed with imagination too shallow to see that great truths can be expressed in poetry and myth."[40]

To sum up the creationist-evolutionist debate as science versus religion is (at best) a gross, misleading, and dangerous oversimplification. This battle is instead one within religion, which has spilled over into the political arena. The Protestant theologian Langdon Gilkey described the fights about the early 1980s Supreme Court case involving Arkansas "equal time for creation" law as a battle between "liberal religion and liberal science on the one side and absolutist religion and its appropriate 'science' on the other."[41] The word "liberal" in Gilkey's description does not refer to political leanings in the sense of Ted Kennedy, but rather pertains to those values of freedom of expression and freedom of inquiry that arose during the Enlightenment. It should be noted that in this case, federal judge William Overton ruled that the teaching of creation science in public schools—even just along side evolution—was in essence the establishment of religion by the government and hence unconstitutional. A few years later, a seven-to-two majority of the Supreme Court ultimately sided with Overton's reasoning. Although some of the words and the players have changed in the past 20 years, today's events replay the same old song. The core of intelligent design, like its intellectual ancestor "creation science," is not science, but religion—an absolutist religion that runs counter to the principles of the Enlightenment. Echoing Overton's decision, Judge Jones, who presided over the Dover case, ruled that the promotion of intelligent design in a science class in public schools amounted to the establishment of religion.

■ The Dover Decision

The voters in Dover, Pennsylvania did not wait for the judge to issue his ruling in the *Kitzmiller v. Dover* case; all eight ID-supporting members of the school board were voted out of office in November 2005,[42] replaced by science-friendly members. On December 20, 2005, Judge Jones weighed in, with the release of a 139-page ruling that concluded that it was "abundantly clear that the Board's ID Policy violates the Establishment Clause. In making this determination, we have addressed the seminal question of whether ID is science. We have concluded that it is not, and moreover that ID cannot uncouple itself from its creationist, and thus religious, antecedents."[43]

According to Jones, ID is not science for three distinct reasons:

> (1) ID violates the centuries-old ground rules of science by invoking and permitting supernatural causation; (2) the argument of irreducible complexity, central to ID, employs the same flawed and illogical contrived dualism that doomed creation science in the 1980s; and (3) ID's negative attacks on evolution have been refuted by the scientific community.[44]

Any one of these reasons would be sufficient to rule that ID is not science, but the judge found three. He also discussed exactly what was wrong with the disclaimer read to the students in Dover.

> In summary, the disclaimer singles out the theory of evolution for special treatment, misrepresents its status in the scientific community, causes students to doubt its validity without scientific justification, presents students with a religious alternative masquerading as a scientific theory, directs them to consult a creationist text as though it were a science resource, and instructs students to forego scientific inquiry in the public school classroom and instead to seek out religious instruction elsewhere.[45]

The trial in Dover was just in district court, and its findings technically are precedent only in that court. Because the new evolution-friendly board declined to appeal Judge Jones' ruling, this case will go no further. The struggle over the teaching of evolution and ID in public schools continues; ID-friendly or anti-evolution legislation is pending in several states, including Maryland, Indiana and a few southern states. Yet the *Kitzmiller v. Dover* decision appears to have legs. The Ohio State Board of Education had included in its standards a passage singling out evolution for "critical analysis." That passage was struck out when Republican Governor Bob Taft questioned its legality in light of the Kitzmiller verdict. A similar backlash against ID has been seen in Wisconsin and Utah, where proposals to introduce ID or anti-evolution legislation were scrubbed.[46]

■ Return to Kansas

There is a certain irony to the perceived notion that Kansas is replete with citizens that are ignorant of or hostile toward evolution. Many of the dinosaur bones discovered in the 1870s that provided early support for evolution were found in this region. The journal *Evolution*, arguably the premier journal of the field, is published in Lawrence, Kansas. The University of Kansas and Kansas State University both support strong research programs in evolution and related disciplines, such as ecology and genomics. As far as I know, the Kansas Board of Education did not solicit input from any of the faculty at these institutions in the drafting of its standards.

In November 2005, shortly before the board approved the new standards, I was in Kansas for a conference on ecological genomics, the study of the genetic basis of ecologically relevant traits (such as longevity or a plant's resistance to insects). Such studies address questions such as, "What are the genetic and regulatory mechanisms involved in how an organism responds to environmental changes? What is the ecological context necessary to understand gene expression?"

One of the speakers at the conference at the mention of evolution (the E word) said in a mock conspiratorial voice, "I hope that the Kansas School Board isn't listening." My thought was that perhaps the board should be listening; that we biologists should be explaining to them about our research and how evolution is an essential component of that vital research, with practical real-world applications as well (see chapter 1).

Biologists need to spend more time explaining why intelligent design is not a credible scientific alternative—indeed, not a scientific theory at all—to the evolutionary theory developed by Darwin and his intellectual descendants, and the rich collection of evolution facts that it explains. But this is not enough! The dearth of discourse regarding evolutionary biology in intellectual life extends beyond creationism; many well-educated people who accept evolution grasp how it works at only the most superficial level, and often hold on to serious misconceptions regarding its action. This lacuna exists in part because scientists have not done enough to explain the centrality of evolutionary theory to understanding both the remarkable adaptations of organisms and their diversity.

In addressing the public, biologists as well as other scientists should be cognizant that the majority of the public is religious and/or spiritual, and that many feel threatened by science. Scientists need to communicate better that science can only address questions about the natural world; it doesn't address the supernatural. Moreover, science does not explain morality. Both scientists and nonscientists should realize that although science can and should inform studies of morality by explaining how the world is, science cannot tell us what ought to be. Nor does science explain such qualities as beauty, wonder, and love. In the words of Rabbi Michael Lerner, "I don't expect science to discover a gene that will explain beauty or tell me why I respond with awe and radical amazement at the grandeur of the universe (although part of what I am responding to is an understanding of the universe and how it operates that I have gained from science)."[47]

■ Recommended Reading

Several excellent books exist on the history of creationism, including its most recent variant: intelligent design.

Larson, E. J. 1997. *Summer for the Gods: The Scopes Trial and America's Continuing Debate over Science and Religion.* Harvard University Press.

Numbers, R. L. 1992. *The Creationists: The Evolution of Scientific Creationism.* University of California Press.

Pennock, R. T. 1999. *Tower of Babel: The Evidence against the New Creationism.* The MIT Press.

Scott, E. C. 2005. *Evolution vs. Creationism: An Introduction.* (paperback ed.) University of California Press.

Shanks, N. 2004. *God, the Devil, and Darwin: A Critique of the Intelligent Design Movement.* Oxford University Press.

Although the Galapagos Islands are forever associated with Darwin and his voyage on the *HMS Beagle*,[1] he actually spent only five weeks of that five-year expedition exploring and collecting in that oceanic island chain; he devoted far more time to the interior of South America. The Galapagos certainly deserves its reputation as a critical site for the formation of Darwin's ideas of evolution, but Darwin did not fully realize the import of his work there until long after he returned to home to England. There, an ornithologist colleague named John Gould informed Darwin that the mockingbirds he had collected on the different Galapagos islands were sufficiently distinct from each other as to be separate but related species, none of which had been observed before. Based on these observations and others, Darwin surmised that the mockingbirds had descended from a common ancestor but had been modified slightly on each island such that they became different species. Soon, Darwin would make similar conclusions about the finches and the tortoises that he had collected in the Galapagos, thus extending the scope of common descent with modification as an explanation for the diversity of life.

More than a century and a half after Darwin's stay in the Galapagos, evolutionary biologists continue to conduct studies in the archipelago. Among other things, these studies have demonstrated the power of Darwin's primary mechanism of evolution—natural selection. Since the early 1970s, the British biologists Peter and Rosemary Grant and their associates have been tracking the species of finches named after Darwin that dwell in these islands.[2] In order to get to know a reasonably sized group of birds intimately, the Grants initially focused their attention on a particular hundred-acre-sized island, Daphne Major, and one species of finch, Darwin's medium ground finch *(Geospiza fortis)*. During what should have been the rainy season of 1977, instead of the typical five or six inches of rain, less than an inch of rain fell on Daphne Major. As a result of this severe drought, seed density dropped dramatically and the composition of seeds changed. Most finches starved to death; the number of the medium ground finches dropped from 1,300 to fewer than 300 during calendar year 1977. But the ones that survived were subtly but noticeably different from those that perished! Although many large birds died and some small birds lived, the survivors *on average* were slightly larger in size and had slightly deeper beaks. These large birds with deeper beaks were better equipped to cope with the new environmental conditions that the drought had produced. Moreover, this trend toward larger birds with deeper beaks persisted in their offspring. Climatic conditions changed again six years later; during the wet season of 1983, over 55 inches of rain fell on Daphne Major. Small, soft seeds increased in abundance. Finches that were smaller with shallower beaks became favored. The Grants also documented several other changes in the morphologies of these birds in response to other climatic changes in the years since.

Rapid evolution isn't just restricted to the Galapagos, or to islands; it can and does happen anywhere. Consider *Drosophila subobscura*, a distant cousin of the well-known fruit fly *Drosophila melanogaster*. Native to Europe, this fly was accidentally introduced in North America in the early 1980s and was first found in Port Townsend, Washington. It has since spread southward to central California and northward to southern British Columbia. In 1997 and 1998, George Gilchrist and Ray Huey, two biologists then at the University of Washington, collected these flies from various locations across their range and then reared their offspring in a common lab environment.[3] They found that offspring of the flies collected in northern regions had wings that were larger than and were shaped differently from the ones from the offspring of the southern populations. Offspring of flies from the middle of the range were intermediate; indeed, the traits of the populations going north to south just graded into each other without abrupt changes. Biologists call such a pattern of gradual change across a geographic region a cline. This cline in North America formed in less than 20 years and mirrors similar clines seen in European populations of *Drosophila subobscura*.

The pollution caused by the industrial revolution during the late eighteenth and early nineteenth centuries in the developed world has led to many instances of insects becoming darker in color. This phenomenon, known as industrial melanism, happened because as the soot and other pollutants made trees darker, light-colored forms of organisms became more conspicuous to predators. Industrial melanism has been most intensively studied in the peppered moth, *Biston betularia*. In the 1840s, a dark morph of this moth appeared in the United Kingdom; during the next few decades, this morph increased in frequency until it became the most prevalent in the populations surrounding industrial centers in the United Kingdom. Since the more recent passage of "clean air" laws, pollution has decreased and the trees have become lighter. Following this reduction in pollution, the light form of the peppered moth has increased in frequency, such that the dark forms now are becoming increasingly rare. The color difference between forms is genetic, and is probably due to a single gene (or gene complex) with modifying genes. Bruce Grant and Larry Wiseman at the College of William and Mary documented a similar decline in the frequency of dark forms of the subspecies of the peppered moth that is found in the United States.[4]

The changes in the finch's beak, the fly's wing, and the moth's color are examples of evolution *by natural selection*. These and numerous other studies demonstrate that not only does natural selection occur but that it also can lead to very rapid evolution—far faster than Darwin had thought.

Natural selection, as Darwin saw it, is simply the result of a few properties that almost all populations of organisms share. First, in every species, more individuals are produced than can survive and reproduce. This "reproductive excess" leads to what Darwin called "a struggle for existence."[5] Consider all of the acorns produced by a mature oak tree. Only a small fraction of them will ever survive to become seedlings, much less mature oaks. Such reproductive excess is also a feature of less prolific organisms. In fact, Darwin had calculated that even elephants, notorious for their long generation time and small number of offspring per pair, would take over the earth in less than a thousand years if there were no checks to their reproduction and survival. Because the earth is not overrun with elephants, reproductive excess must exist even in that long-lived, slow-reproducing species.[6]

A second requirement for natural selection is variation. Within populations, individuals vary. From a distance, all finches look alike. But if you look closer, subtle differences emerge. One finch is slightly darker than average. Another has a beak just a hair larger than average. Still others have barely noticeable differences in wing shape and size, coloration and shading patterns, and other measurable traits. Careful observation allowed Rosemary and Peter Grant to document the individual variation in these birds in several morphological traits, especially their overall size and their beak shape.

For natural selection to occur, this variation in traits must have fitness consequences. That is, the variation must affect either survival and/or reproduction. If having a white patch of feathers is just as good in terms of survival or reproduction as a red patch, patch color is not under natural selection. As we will see in chapter 3, many DNA changes exert little or no effect on survival or reproduction. In the population of Darwin's finches that they studied, however, the Grants found that morphological differences did affect survival, at least at certain times. Birds with deeper beaks were better able to survive the drought because they were more proficient in acquiring the

Figure S.1 Illustration of variation in one species of Darwin's finches, the medium ground finch (*Geospiza fortis*). Taken at the Smithsonian Museum in 2005. Courtesy of Sarah Huber.

type of seeds that had grown despite the drought. Hence natural selection, under drought conditions, favored the deeper beaks.

Evolution by natural selection requires one more condition: those traits that affect fitness must be passed on to the next generation. Darwin knew nothing about DNA; he didn't even know about Mendel's laws of genetics. Despite not knowing the mechanisms by which heredity worked, Darwin knew from observation and his work as a breeder (especially of pigeons) that individuals tend to resemble their parents. Parents that are tall tend to produce offspring that are taller than average. This resemblance between parents and offspring is what allows populations to respond to natural selection and exhibit evolutionary change. The genetic variants that gave a deeper beak shape and a larger body size to those Darwin's finches that survived the drought were passed on to their offspring. As a consequence, the offspring were more like their parents than they were like the original population.

Evolution and natural selection are not the same. Evolution can take place without natural selection. In fact, as we will see in chapter 3, much evolutionary change at the DNA level is due to mutation and the random process of genetic drift. Conversely, natural selection does not necessarily lead to evolutionary change. Natural selection can help maintain the status quo; many more mutations are deleterious than are beneficial, and selection acts to weed out the bad mutations. This "weeding out" type of natural selection is called negative selection to contrast it with the type of selection (positive selection) that results in rare, beneficial genetic variants becoming more common. It is this positive selection that is required for the changes seen in the wings of *Drosophila subobscura* and the beaks of Darwin's finches. In chapter 5, we will encounter a third type of natural selection called balancing selection, which actively preserves genetic variants within populations.

As we've seen in the above cases, scientists sometimes can, by careful and painstaking observation, directly observe evolution by natural selection. Sometimes they can catch it in the act. But what about evolution by natural selection that occurs at rates much slower than the episodes of selection scientists can observe? Given the immense age of the earth, as Darwin realized, even changes very slow by human standards can be blinks of an eye from a geological standard. A 1% increase in body size over a thousand years is far too slow to be detected. That change, maintained and compounded over just a quarter of a million years, would transform a mouse that weighs an ounce into an animal that weighs almost a pound—the size of a rat. In another quarter million years, that mouse would be the size of a house cat, tipping the scale at nine pounds. Obviously, such sustained changes do not usually persist for such long periods; but when they do, the results are staggering.

Evolutionary geneticists are now able to detect the action of natural selection by making inferences from the patterns of changes seen in the DNA. Natural selection that has operated in the past has left the equivalent of footprints. To detect these footprints, evolutionary geneticists rely upon DNA data (taken from many different individuals, usually in more than one species), statistics, and mathematical theory. We turn to such studies in chapters 4 and 5.

3 ∎

Negative Selection and the Neutral Theory
of Molecular Evolution

> *Production of man from a primitive jawless fish in half a*
> *billion years is a remarkable example of progressive evolution*
> *but we should not forget that degeneration and extinction are*
> *much more common in evolution.*

—Mooto Kimura (Kimura, 1983, p. 61)

∎ The Irony of Molecular Evolution

One irony of molecular evolution is that the genes that are most important to the function of the organism usually have changed least through evolutionary time. Consider histone H4, one of a number of proteins known as histones that binds to DNA and allows for the proper expression of genes.[1] The sequence in histone H4 in pea plants differs from that in mammals by only two changes in a total of 102 amino acid sites. Only two changes occurred between animals and plants! Assuming that plants and animals last shared a common ancestor 1.2 billion years ago, this is an evolutionary rate of 0.008 changes per site for each billion years.[2]

Contrast this strong conservation of the histone H4 protein with the comparatively rapid evolution of fibrinopeptides. In the last chapter, we saw that these are parts of fibrinogen that are cleaved off by thrombin to produce fibrin in the blood-clotting process. They have little function afterward. These fibrinopeptides evolve about a thousand-fold faster than the histone H4.

Why should these almost-dispensable fibrinopeptides evolve so much faster than the essential histone H4? This seemingly paradoxical finding that functionally less important genes evolve faster than those with more important functions makes sense when we consider that far more of the mutations that have an effect on an organism are harmful than are advantageous. Why is this so? Consider a well-functioning watch. Banging that watch on a table is more likely to result in worse rather than better performance of the watch. The same is true for well-functioning organisms. Change is more likely to make things worse than better. Evolution seems to also hold to the principle, "If it ain't broke, don't fix it." One the one hand, proper blood clotting is not very dependent upon the exact nature of the amino acid sequences of the fibrinopeptides as long as they prevent fibrinogen molecules from sticking to each other so that they can be cleaved off by thrombin under the appropriate conditions. On the other hand, the exact amino acid sequence of histone H4 is essential to gene expression and hence to the viability of the organism. Thus, the DNA sequence that produces fibrinopeptides can tolerate mutations far better than the DNA that produces histone H4.

Because many more harmful than beneficial mutations occur each generation, natural selection usually acts as the reaper, weeding out those deleterious mutations. This weeding type of selection, called negative selection, is much more common than positive selection, the form of selection that increases the frequencies of advantageous genetic variants.

Of course, advantageous mutations do occur sometimes, and positive selection does operate. Adaptations such as the dog's keen sense of smell, the antelope's fast gait, and the cactus plant's waxy cuticle (which enables it to survive extreme drought) are the result of new, advantageous mutations rising in frequency via positive natural selection. Further, new advantageous mutations are required for adapting to new climates, such as what happens when a species expands its range. Mutations also are required for the process of coevolution, when different species respond to one another. An example of such as process of coevolution occurs when a predator species and its prey are engaged in an "arms race"; the prey adapt to changes in the predator and vice versa. Similar coevolutionary arms races occur when a species evolves resistance to a disease and the disease evolves counter-measures to the resistance. Despite positive selection being essential for adaptive evolution, this form of selection is considerably rarer than negative selection.

The conservation of functionally important DNA sequences due to the operation of negative selection also has practical importance. With the ever-increasing number of DNA sequences in databases, more and more biologists are using this principle of conservation to identify new genes and make inferences about their presumptive function. If a DNA sequence from a sea urchin closely resembles a gene found in mammals that produces a protein involved in transporting substances across cell membranes, then this new sea urchin sequence is likely to also play a role in membrane transport.

In the next chapter, we will examine how molecular evolutionary biologists developed tests to detect the action of positive selection operating on particular genes. These tests rely, however, on a deep understanding of how evolution would work in the absence of positive selection. A Japanese mathematical geneticist named Motoo Kimura developed such understanding in the late 1960s; in the decades since, his theory has revolutionized the study of molecular evolution.

■ Kimura's Revolutionary Idea

The year 1968 was one of worldwide change and political turmoil. It was the year of the Tet Offensive, the assassinations of Martin Luther King and Robert F. Kennedy, riots in the streets of Chicago during the Democratic convention, and the Prague Spring that ended with the Soviet Union's invasion of Czechoslovakia. In that same year, Motoo Kimura put forth a radical theory called the "neutral theory of molecular evolution" that would revolutionize the field of evolution.

In 1968, Kimura, at age 43, was arguably the leading mathematical evolutionary geneticist of his generation.[3] Born and raised in Japan, Kimura had received his Ph.D. at the University of Wisconsin in Madison a decade before under the tutelage of James Crow. (In January 2006, Crow celebrated his ninetieth birthday, still much engaged in science as the elder statesman of evolutionary genetics.) In the 1950s and 1960s, Crow had been studying effects of mutation in fruit flies; he found many more mutations with subtle effects on viability and reproduction than mutations that killed the flies. Kimura acquired from Crow an appreciation of the evolutionary importance of mutation and negative selection.

Kimura also had been interested in the new data coming from molecular genetics, namely protein sequence data. Although 1968 was 15 years after Watson and Crick presented the model for the structure of DNA, biologists were still unable to get large amounts of information from the genetic material; techniques for sequencing more than a handful of nucleotides of DNA at a time would take another decade. Biologists were, however, able to obtain the amino acid sequences from proteins, the best available alternative to the gene itself. During the 1960s, more and more sequences of the amino acids of proteins from a variety of organisms were becoming available. Kimura had been looking at the rates of changes in amino acid sequences from the few proteins that had been sequenced at the time. The rates of change at individual proteins were not very fast, but Kimura realized that genomes were vast. Kimura extrapolated the rates of evolution observed in a few proteins to the whole genome, and estimated that in mammals a substitution of one genetic variant for another occurred every other year. If so many substitutions were taking place, could positive selection really be driving most of them?

Kimura was also impressed by the pattern of gene substitution called "the molecular clock." The double Nobel laureate Linus Pauling and others showed that amino acid sequences of proteins seemed to evolve at a constant rate in different types of organisms. Moreover, for any given protein, the rates seemed to be constant with time; a protein that changed by 10% between two species known to be separated by 40 million years of evolutionary divergence generally changed by about 20% between two species known to be separated by 80 million years. Recall that not all proteins evolve at the same rate. In fact, substantial variation exists in their rates of evolution; for example, fibrinopeptides evolve hundreds of times faster than histone H4. Thus, we see that molecular clocks based on different genes run at a variety of rates.

Pauling and others saw that the concept of the molecular clock would be a great boon for evolutionary studies by allowing one to estimate, using solely molecular data, the point at which two species diverged from a common ancestor. Given the example of the proteins mentioned above, the molecular clock predicts that two species differing by 5% in the sequence of that protein would probably have diverged about 10 million years ago.

Why should the rates of molecular change be more or less the same in different parts of the evolutionary tree? We know that rates of morphological change can vary a great deal. For instance, primate morphology seems to change faster than that of amphibians. Although humans look quite different from chimpanzees, many species of frogs that diverged much longer ago than did humans and chimpanzees look very similar to each other. Modern horseshoe crabs haven't changed much in morphology from their 200-million-year-old fossil relatives.

Kimura's proposition was that at the molecular level, the vast majority of new mutations are either deleterious or neutral with respect to selection.[4] Negative selection would quickly weed out the deleterious mutations. Neutral variants, however, would be subject to chance sampling every generation, a random process that evolutionary geneticists call genetic drift. Although nearly all newly arisen, neutral mutations would be lost from the population after a few generations, a lucky few would rise in frequency. Occasionally, one very lucky variant would eliminate all others in the population even if it was no more fit or less fit than the other variants. Evolutionary geneticists refer to such a variant that rises to 100% frequency as being fixed in the population. Note that if one genetic variant at a site becomes fixed, all other variants at that site must have been eliminated. (The terms "fixation" and "substitution" will be used interchangeably throughout the book.) Advantageous variants become fixed via natural selection, but the fixation of neutral variants through the lottery of genetic drift occurs due to chance, not selection.

Most evolutionary biologists, at first, viewed Kimura's neutral theory with a great deal of skepticism, if not hostility.[5] To some, Kimura's ideas seemed to be counter to Darwin's and the mainstream evolutionary thought of the 1960s; indeed, the year after Kimura's paper appeared, two biochemists, Jack King and Thomas Jukes, published a proposal similar to Kimura's with the title "Non-Darwinian Evolution."[6] The name "non-Darwinian evolution" for these ideas dropped out of favor because evolutionary biologists realized that Darwin himself did not view natural selection as the exclusive means of evolutionary change and had discussed the role of chance processes in evolution. Kimura's term—the neutral theory of molecular evolution—fared much better.

During the 1960s, the leaders in evolutionary biology had focused on positive selection and adaptation, almost to the complete exclusion of other evolutionary forces. For example, in 1963 Ernst Mayr, one of the domineering figures in evolutionary biology, wrote that it was "exceedingly unlikely that any gene will remain selectively neutral for any length of time."[7]

Population geneticists had long been aware of genetic drift as an evolutionary force that could alter frequencies of genetic variants in small populations. An isolated group of a few individuals would have different frequencies of genetic variants than the main population from which it derived. One consequence of such founder effects is that rare diseases are sometimes more common in human populations that were derived from a small number of individuals. Some examples include Tay Sachs disease in Ashkenazi Jews and congenital colorblindness in peoples of the Pingelap Atoll in Micronesia. Sewall Wright, one of the founders of the fusion of Mendelian genetics and Darwinian evolution during the 1920s and 1930s known as the modern synthesis, had championed the role of genetic drift in evolution. Wright's notion was that in situations in which a species was found in many small, semi-isolated populations, genetic drift would cause different local populations to have different frequencies of genetic variants from each other. These local differences would allow populations "to test out" different combinations of genetic variants. Some of these local populations would hit upon really good combinations, and selection among those local populations would cause those superior combinations of variants to become fixed within the big population. Aspects of Wright's theory remain controversial to this day.[8]

Many population geneticists prior to Kimura dismissed the importance of genetic drift as an evolutionary force because its strength decreased with increasing population size. Genetic drift, as a sampling process, is like tossing coins. A coin—unless it is rigged—should land heads up half the time and tails up half the time. But that's just the expected result; it's not unusual that seven of ten coins would land heads. It would be unusual if 70 of 100 coins landed as heads. If 700 of 1,000 coins were heads, you would be convinced that the coin had been tampered with. Genetic drift follows similar patterns of probability. In a small population, genetic drift can easily alter the frequencies of genetic variants. This chance process can allow even some deleterious variants to become fixed in the population. In larger populations, however, genetic drift is less potent relative to natural selection, and natural selection can act on variants that have even relatively small differences in fitness. Small populations, on the other hand, are less likely to persist for long periods of time.

One revolutionary aspect of Kimura's proposition was his view of positive selection as an insignificant force in molecular evolution. Even Wright, the champion of genetic drift, was not ready for that. Positive selection, both within and among populations, was an essential component to Wright's theory. In contrast, in Kimura's view of molecular evolution, positive selection is so rare that that it doesn't affect the observed patterns of molecular evolution.

Kimura's ideas were revolutionary also because he argued that the evolution of molecules is (largely) decoupled from what we ordinarily think of as evolution—changes in morphological, physiological, developmental, and behavioral traits. To Kimura, these traits—what biologists call phenotypic traits—were largely distinct from molecular traits. Kimura didn't deny that phenotypic evolution traces to changes in DNA sequences, or that adaptations

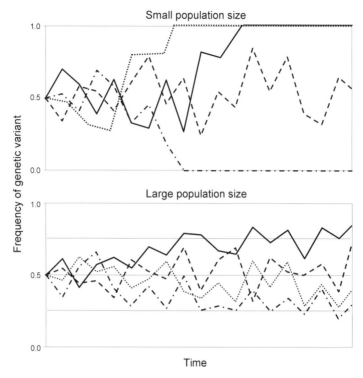

Figure 3.1

Genetic drift is more potent in small rather than large populations. Horizontal and vertical axes represent time and frequency, respectively. The individual lines represent the frequency of a specific genetic variant for hypothetical repeated examples of replicate populations of the same size; all populations began with the same initial frequencies of the genetic variant. Note that some variants went to fixation or loss in the small populations but none did in the large populations.

such as the sonar-like navigational abilities of bats and the elaborate dances of bees to communicate nectar stores are the products of positive selection. Instead, Kimura noted that these changes in DNA that affect phenotypic traits are only a very small fraction of the overall changes in DNA sequences. Thus, the changes caused by positive selection on phenotypic traits could have a marginal effect on the overall patterns of molecular evolution. In a sense, Kimura was like Einstein, who in his theory of special relativity limited the scope of Newton's laws of motion; Newtonian physics worked well for sufficiently low velocities but failed when velocities were appreciable fractions of the speed of light. Although Kimura allowed for pervasive positive selection acting on phenotypic traits, he concluded that genetic drift, mutation, and negative selection overshadow positive selection in the molecular world.

■ **Support for the Neutral Theory**

Some of the strongest support for the neutral theory comes from genes that have lost their function and are unable to produce proteins. Such genes without function, known as pseudogenes, should not experience negative selection. If negative selection were more important than positive selection, we would expect these pseudogenes to have very fast rates of evolution. The rates of evolution for pseudogenes were not known when Kimura first proposed the neutral theory, so such data would be strong "after the fact" evidence for or against the neutral theory. In the early 1980s, molecular evolutionists determined that the rate of pseudogene evolution is very high, among the most rapid rates of evolution—consistent with Kimura's prediction.[9]

Kimura used his neutral theory to provide theoretical justification for the molecular clock, the relative constancy of evolutionary rates across different organisms in a particular protein or DNA sequence. The key point of Kimura's justification is that the rate that new variants become fixed in populations (the substitution rate) is exactly equal to the rate at which neutral mutations appear. Although we will take this supposition as a given for now, the rationale behind this simple relationship is explained in the box below. Provided that the neutral mutation rate does not change much among different organisms, the rate of molecular evolution should be constant. Two factors determine the neutral mutation rate: the overall mutation rate and the extent of functional constraint (the fraction of mutations that are neutral versus those that are deleterious). If these factors remain more or less the same, then the neutral mutation rate will be relatively constant. All other factors—for instance, the population size or the amount that the environment fluctuates—are largely irrelevant to the rate of molecular evolution.

■ **Why the Substitution Rate Equals the Neutral Mutation Rate**

Why should the substitution rate be equal to the rate at which neutral mutations arise? The rate of substitution would equal the number of total mutations each generation multiplied by the probability that any one of those mutations will eventually become fixed in the population.

Let's start with the first part of the equation—the total number of mutations. Every generation, new mutations appear in the population at a rate of $2N$ times the mutation rate, where N is the population size. Large populations contain more individuals and thus, more opportunities for mutation. The "2" comes in because, in most species, individuals receive two distinct sets of genes—one from their mother and one from their father. Mutation rates are tiny. The typical gene that codes for a protein has a mutation rate on the order of around one in a million. A nucleotide

■ Why the Substitution Rate Equals the Neutral
Mutation Rate *(continued)*

within a gene has a mutation rate on the order of one in a billion (one-one thousandth of one in a million). Population sizes, however, can be rather large but variable. Species smaller than about 1,000 are in critical danger of going extinct. Population sizes can number in the billions, however, particularly for species with small body sizes (for example, insects, most fungi, and bacteria). Therefore, the mutation rate times the population size generally yields a modest number.

Recall that the neutral theory postulates that virtually all mutations are either deleterious or essentially neutral. The deleterious mutations don't persist, so when considering the substitution rate, we only consider the neutral mutations. Note that the neutral mutation rate is the overall mutation rate multiplied by the fraction of mutations that are neutral. According to the neutral theory, all genetic variants persisting in a population have an equal probability of being the one that becomes fixed in the population. Thus the probability of a variant becoming the one fixed in the population at any given time is the same as its current frequency. If a hypothetical variant of a gene is present at 34% in the population, it has a 34% chance of becoming the variant that is fixed. This makes intuitive sense—the variants all are equally fit, and none is more or less likely to be the lucky one than any of the others. It's like a lottery; your chances of winning depend upon how many tickets you possess. The new variant is unique; it is the only one of its kind. Call the number of individuals in the population N. Because every individual has two copies of genes, two times N copies of genes are in the population. The new mutation thus has a one in $2N$ chance of being the lucky variant that wins the genetic lottery and becomes fixed in the population.

To figure out the rate of substitutions, we need to multiply how many new mutations occur and the probability that any one new variant will be the lucky one.

$2N$ times the neutral mutation rate of new mutations every generation multiplied by $1/2N$ probability of being lucky equals the neutral mutation rate.

An interesting thing happens; the two $2N$s cancel each other out, making the math really simple. The rate of substitution is thus equal to the neutral mutation rate. Of course, this is the *neutral* mutation rate. If negative selection is very strong, then the neutral mutation rate may be very much lower than the actual mutation rate. In the case of amino acids in histone H4, the neutral mutation rate may be around one-thousandth of the actual mutation rate.

At the time Kimura had published his first paper on neutral theory, evolutionary biologists were just starting to use a technique called protein electrophoresis to measure the extent of variation in proteins within natural populations in a variety of organisms. Electrophoresis distinguishes protein variants based on their ability to move across an electrically charged, starchy gel. Variants of proteins that differ in charge or size owing to slight differences in their amino acid sequence will move down the gel at slightly different rates. These variants can thus be easily visualized. The virtue of this technique is that many individuals can be quickly assessed for differences at a couple of dozen different proteins. Not only was the technique easy to use and relatively inexpensive, but it also could be used in natural populations in just about any organism. If you could find 'em and grind 'em (to get out their proteins), you could study the variation in their proteins. Moreover, protein electrophoresis requires only a tissue sample, not the whole organism; in most cases, one can take this sample without harming the organism. Due to its ease, the technique of electrophoresis became a bandwagon within evolutionary biology circles during the 1960s and especially 1970s. In most species, populations were found to be surprisingly variable. In organisms ranging from flies to humans, over 30% of the proteins sampled were variable.[10] Moreover, often up to about 10% of individuals had two different variants for a particular protein. Such individuals, called heterozygotes, received one variant from their mother and one variant from their father. How could so much genetic variation exist in populations? Could all this variation be maintained by natural selection?

The neutral theory also makes predictions about the extent of polymorphism within species. One of Kimura's major insights was seeing that polymorphism was the transient stage in the turnover of neutral variants. Polymorphism and evolutionary change were just two sides of the same coin. The extent of polymorphism should thus be proportional to the length of time individual variants took on their journey to becoming fixed or eliminated. Consider genetic variants whose frequency is primarily influenced by drift; such variants will spend more time before either becoming fixed or going extinct in larger populations than they do in smaller populations. As an analogy, suppose that Joan and James are both flipping coins, and Joan flips 10 coins at a time while James tosses 100 coins at a time. Simply because she flips fewer coins at a time, Joan will almost certainly throw one set with all heads or all tails long before James ever does. The same principle works with genetic variants controlled by genetic drift; these variants will be lost or fixed in smaller populations much faster than they will in large ones.

The census population size is not the only factor that determines the effect of genetic drift. If many more females than males occur in a population, genetic drift will be stronger than one would expect based just on the total numbers of offspring. Genetic drift is a sampling process that occurs in the making of gametes and the fertilization of those gametes into zygotes; males,

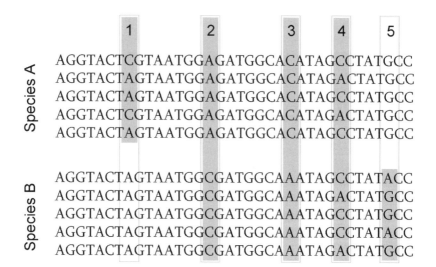

Figure 3.2

Polymorphism and divergence in species. Each row represents the DNA sequence of a particular gene from one chromosome of individuals from two species A (top) and B (below). At site 1, a *polymorphism* (for nucleotides A and C) exists within species A but not species B (all sequences are A). Sites 2 and 3 are examples of *divergence;* all sequences from species A differ from all sequences of species B at this site. Site 4 is an example of a *shared polymorphism;* some sequences are C at this site and some are A. At site 5, species B is *polymorphic* (A and G) and species A is *monomorphic*. All other sites are monomorphic.

as the rarer sex, represent a genetic bottleneck. The same principle would hold if females were rarer than males. A population of 100 males and 100 females is affected less by genetic drift than either a population of 50 males and 150 females or one with 150 males and 50 females. Temporal fluctuations in population size also influence the effect of genetic drift. Suppose a population regularly cycles from 100 in one generation to 1,000 in the next to 10,000 in the third. This population with fluctuating population sizes experiences a greater effect of genetic drift than a population whose size remains constant at 500. Furthermore, if not all individuals breed, then the effect of genetic drift would be higher than expected based on the census population.

Sewall Wright introduced the term "effective population size" to serve as a measure of how much genetic drift any given population experienced.[11] A population that has an effective population size of 1,000 would experience genetic drift in the same way that an idealized, randomly mating population of equal numbers of males and females would that was stable at 1,000 breeding individuals in each generation. This measure of "effective population size" allows for comparisons among populations and species with very different demographic properties. We will encounter this measure several times throughout the following chapters.

The neutral theory also predicts that the neutral mutation rate will affect genetic variability. Mutations yield new genetic variants, thus an increased mutation rate would also increase the amount of polymorphism. Because mutation rates generally vary much less than do effective population sizes, population geneticists usually focus more on the latter.

Species with larger population sizes do tend to exhibit more genetic variability. Bacterial populations have more variation than do most insect species, and most insect species contain more variation than do most vertebrate species. The effective population size of humans is rather low, despite their current large census population. This apparent anomaly is almost certainly due to low historical population sizes of humans, and effective population size measures being sensitive to dips in population sizes. We will consider the effective population size of humans in more detail in chapter 6. At the extremes, elephant seals and cheetahs both have recently undergone narrow population bottlenecks in the recent past, and both species exhibit exceedingly little genetic variation.

■ Kimura's Legacy (Part 1)

The protein data for divergence among species and polymorphism within species are thus at least roughly consistent with the neutral theory. But this doesn't prove that the neutral theory is correct. As some evolutionary biologists argued, these results also could be consistent with models of evolution that assume some forms of positive selection. During the 1970s, many evolutionary biologists took stands in what has been called the selectionist-neutralist debate. The neutralists defended the neutral theory whereas the selectionists argued for a greater role for positive selection. Although this debate still simmers, most participants have taken more nuanced stances.

In 1991, Ernst Mayr admitted that his initial position on the neutral theory was incorrect:

> Although this theory was at first vigorously opposed by most evolutionists, including myself, the high frequency of "neutral" base-pair replacements is now well established. On the other hand, the selective significance of numerous alleles that had been considered neutral by neutrality enthusiasts has also been established.[12]

In the next chapter, we will further discuss the legacy of Kimura and the neutral theory. Today, virtually all evolutionary biologists see that Kimura's neutral theory plays a very important role in tests for detecting positive selection. To understand this role of the neutral theory, we must discuss how scientists test hypotheses.

■ The Neutral Theory, Hypothesis Testing, and the Burden of Proof

Practitioners of both legal and scientific reasoning encounter issues pertaining to the burden of proof. In the American legal system, a defendant in a criminal case must be found guilty beyond a reasonable doubt in order to be convicted. That is, the burden of proof is on the prosecution rather than the defense to make its case. Based on English common law, the Anglo-American legal system puts the onus on the prosecution to prove guilt because we believe that it is better to let a guilty person go free than it is to imprison an innocent person. The burden of proof is much lower in civil cases, however. In civil cases, the plaintiff's side need not prove their case beyond a reasonable doubt. The O. J. Simpson cases illustrate this distinction; although Simpson was found not guilty of murder in a criminal court, he lost the wrongful death suit in civil court.

Discussions about the burden of proof also appear in science. Given two different explanations that are both consistent with the observations at hand, which should we choose? Scientists and philosophers have been guided by a principle known as Occam's razor, named after the fourteenth-century philosopher William of Occam. According to Occam's razor, when competing explanations or hypotheses can each explain the phenomena equally well, then the one we should pick is the one the one that is least complicated, or the most parsimonious. As an absurd example, suppose that you see a felled tree that was previously standing, and you know that a lightening storm occurred the previous night. One hypothesis is that lightening caused the tree to fall. Another possibility is that a film crew felled the tree, making it appear as if it were struck by lightening. In the absence of prior knowledge of a film crew or ancillary information, the former hypothesis is obviously the most parsimonious. Sometimes distinguishing the most parsimonious hypothesis is not as cut and dry. One phrasing of Occam's razor is "when you hear hoofbeats, think of horses not zebras." Because horses are more common than zebras in America and Europe, horses would be the more parsimonious cause for the hoofbeats in those continents. But in parts of Africa, zebras are more common than horses, and thus zebras would be the more parsimonious explanation there.

So the burden of proof is placed on the more complicated, less parsimonious hypothesis. For it to be accepted, it must have some greater explanatory power than the more parsimonious one. Scientists often frame hypothesis testing by setting up two hypotheses; one is a "null hypothesis," and the other is the "alternative hypothesis." The null hypothesis is typically the hypothesis of "no difference." Take the case of testing the efficacy of a cholesterol-lowering drug. A null hypothesis would be that no real difference would be apparent in the cholesterol levels between those patients given the drug and those given a placebo, that the differences observed are just due to chance. An alternative hypothesis would be that the patients who received the drug had lowered their cholesterol levels more than had those patients who received the placebo. The burden of proof is on the alternative hypothesis; a real difference must exist between the two groups of patients—one that is unlikely to have occurred by chance alone. To take an absurd case, suppose that only one patient was in each of the control and the experimental groups; the patient given the drug reduced his cholesterol by 20 points, while the patient given the placebo reduced his cholesterol by only 12 points. Clearly, the difference could easily arise for chance reasons totally unrelated to the drug. If 40 people had been in each group, however, and all in the drug treatment had reduced their cholesterol levels by more points than all in the placebo treatment, then we could be very sure that the differences were real (due to the drug). The alternative hypothesis met the burden of proof; thus, we could reject the null hypothesis of no difference between the drug and the placebo treatment. Note that failure to reject the null hypothesis doesn't mean that we accept the null hypothesis. In the first case, in which only one person had been in each treatment group, that difference could have been real. We just lacked proof that it was. The alternative hypothesis did not meet the burden of proof.

Biologists, like many other researchers, rely on statistical tests to determine whether differences between groups represent differences that are unlikely to have arisen by chance alone. Two types of errors can be made in these tests. The first is rejecting the null hypothesis—saying that there is a difference between the groups when in fact, no real difference exists. Sometimes, these errors are called "false positives" (note that positive here doesn't mean positive selection). The other type of error is not rejecting the null hypothesis when there really is a difference. These errors are sometimes called "false negatives." By convention, scientists are willing to accept a 5% rate of making false positives. That is to say, the convention allows that the null hypothesis is to be rejected when the probability that the differences among the groups are due to chance is less than 5%. (This level of 5% is somewhat arbitrary and can be adjusted.) Differences that are unlikely to come about by chance alone are sometimes called statistically significant. Differences that are extremely unlikely to come about by chance are sometimes called highly significant. Note that for any particular result deemed statistically significant, a small probability exists that it is a false positive. If many tests are performed, chances are good that at least one of the statistically significant results is a false positive. Further testing can mitigate that problem.

An important property of a statistical test is its ability to reject the null hypothesis—that is, not to make a "false negative" error—at a given level of making "false positive" errors. This property, known as power, is a measure of the discriminatory ability of the test, and it generally increases with increasing the number of data. For instance, the likelihood of making a false negative is lower if one tested 500 individuals with a placebo and 500 with the drug than it would be if one only tested 100 from each group. Testing with the larger samples allows for greater power. For a given number of data, certain tests and experimental designs have more power than do others.

A major strength of the neutral theory is its ability to generate null hypotheses. These null hypotheses consider mutation, genetic drift, and negative selection. Mutation is a fact of life. DNA replication isn't perfect and thus genes can be expected to mutate. Genetic drift is also a necessary consequence of sampling every generation in finite populations. Because most mutations that have an effect are deleterious, negative selection will be pervasive. Therefore, these forces are accounted for in the generation of the null hypotheses. The burden of proof is then placed on demonstrating positive selection. As we will see in the next chapters, that burden has been met for many genes, thus we can infer that positive selection has operated on them.

4 ∎

Detecting Positive Selection

One of the primary goals of population genetics has been to measure and to understand the role of natural selection in shaping variation within and between species. Now that molecular technologies allow genetic variation to be assayed with relative ease, this goal seems within reach.

—John Wakeley (Wakeley, 2003, p. 411)

∎ The Fish That Came into the Cold

The coldest waters in the world are those of the Antarctic Ocean. Here, water temperatures are often just below the freezing point of pure water. High salt concentration allows the water to remain liquid at temperatures where it is ordinarily frozen solid. How do the denizens of these waters cope with such severe conditions? Surely, they must have evolved various specialized adaptations to the frigid water. These waters weren't always frigid; 55 million years ago these waters were rather balmy with average temperatures in the upper 60s, a little chilly for swimming but certainly not bone-chilling cold. The slow cooling of the Antarctic Ocean left its inhabitants three choices: evolve adaptations to the cold, move, or die. Around 25 million years ago, a barrier to currents, called the Polar Front, formed, effectively shutting off the migration option for many species of fish. Although many species went extinct, some did evolve adaptations to the changing climate.

Among the adaptations that some cold-dwelling fish evolved are natural antifreezes. Being less salty than seawater, the blood of these fish ordinarily would freeze. Natural antifreezes, consisting of long repeating chains of amino acids with a sugar periodically attached to one of the amino acids, protect the fish by inhibiting the formation of ice crystals. Molecule per molecule, these natural antifreeze molecules can be hundreds of times more effective than the antifreeze used in cars.

Other adaptations to the cold include changes in the proteins that carry oxygen through the circulatory system, known as hemoglobins (see also chapters 2 and 3). At low temperatures, hemoglobin usually doesn't work well. At the extreme, the icefish dispensed with hemoglobin altogether, and are the only known vertebrates who lack hemoglobin. How do these fish transport oxygen throughout the body? First, the extremely cold temperatures reduce metabolic needs and thus the amount of oxygen that must be transported to the tissues. Yet some oxygen is still needed. Ordinarily, blood plasma is extremely bad at transporting oxygen, but at low temperatures, oxygen can be carried to a limited extent by the blood plasma alone. These fish make up for the low efficiency of plasma transport of oxygen by increasing the volume of plasma and the amount that is moved through the circulatory system in a given time. As a consequence of the extra blood volume, their hearts have become greatly enlarged.

A close relative of the icefish, the Antarctic dragonfish (*Gymnodraco acuticeps*) also made drastic changes in response to cold adaptation but did not go quite as far as its cousins. The dragonfish have hemoglobin, but unlike most vertebrates, in which several forms of hemoglobin are found, these fish possess but a single hemoglobin—one that is specialized for the cold environs.[1] Although the dragonfish hemoglobin differs in many respects from that of most other species, the exact nature of the adaptive role of many of these differences remains a mystery.

In most species, hemoglobin will release oxygen and pick up carbon dioxide more readily when in acidic conditions.[2] What purpose does this response have? Acidity is an indicator that carbon dioxide concentrations are high. Think about soda water, which is acidic because of the carbon dioxide that has been dissolved in it. Thus, the response of releasing oxygen under acidic conditions is usually adaptive because those acidic conditions signal high carbon dioxide concentrations. It would thus be advantageous for hemoglobin to release its oxygen and pick up carbon dioxide in acidic conditions. The hemoglobin of the dragonfish lacks this response; it releases oxygen at the same rate whether in acidic conditions or not.

Biologists are not totally certain about what advantage the lack of response to acidity would give these fish. Perhaps at extremely low temperatures, holding on to oxygen is more important for this hemoglobin than grabbing carbon dioxide, even when carbon dioxide levels are high. Another possibility is that the loss of this response is just the result of a release from negative selection. After all, the icefish totally lost their hemoglobin.

Can we determine whether the changes that led to the dragonfish's type of hemoglobin were driven by positive selection and not simply by a release from negative selection? Will we be able to see whether positive natural selection for cold-tolerant hemoglobin has left traces that can be detected from observing the DNA for the hemoglobin? As we discussed in the previous chapter, to be able to claim that positive selection has been responsible for patterns of DNA changes, first we must rule out explanations based solely on mutation rate, genetic drift, and the extent of negative selection.

Some of the simplest tests for detecting positive selection arise from one feature of the genetic code: some changes in DNA lead to changes in protein, but others do not. The *transcription* of DNA information into RNA information is a one-to-one mapping; that is, for every nucleotide of DNA, there is a nucleotide of RNA. In fact, the only difference is that U (uracil) is used in RNA instead of T (thymine) in DNA. In contrast, the *translation* of RNA information into amino acid (protein) information is not a one-to-one mapping. The information from a group of three nucleotides (called a codon) specifies which amino acid will be in the protein translated from the RNA. There are 64 (4 times 4 times 4) possible codons, three of which are used to indicate when translation should end. So aside from these "stop codons," 61 codons specify the 20 amino acids commonly used in proteins. Thus an average of three different codons code for the same amino acid. Typically, the codons that code for the same amino acid differ in their third codon. For example, the codon UUU codes for the amino acid phenylalanine, but so does UUC. Mutations that do not change the amino acid sequence of the protein

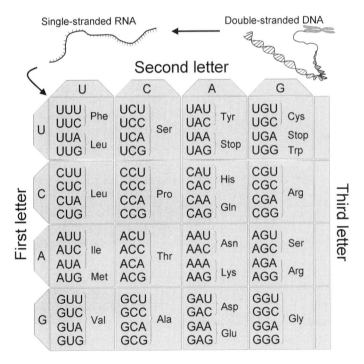

Figure 4.1
The genetic code. Amino acid information is specified from groups of three RNA nucleotides (triplet), which had been specified by DNA nucleotide information.

are called "silent" mutations, and mutations that result in a different amino acid being placed in the chain are called "replacement" mutations. Because they do not alter the nature of the protein, silent changes should affect characteristics of the individual far less than would replacement changes. Thus, on average, replacement changes would be more subject to natural selection (both positive and negative) than would silent changes.

The most straightforward signal of positive selection for new advantageous mutations operating on a gene is a greater rate of replacement substitutions than silent substitutions. Although silent sites often are under some degree of negative selection, it is extremely unlikely that negative selection will affect silent changes more than it will affect changes that alter the amino acid sequence of the protein. The silent sites and the replacement sites are also within the same gene, so they should experience the same levels of genetic drift. Thus, if the replacement sites are evolving faster than the silent changes, the most likely cause is positive selection. Such a test would need to take into account the ability for each site to make silent versus replacement changes. If it is easier to mutate to a replacement site than a silent site, then a faster rate of replacement site evolution may just be an artifact. Such tests also need to rule out chance as the reason for an elevated rate.

Let's return to the Antarctic dragonfish and its hemoglobin. After taking into account how often silent site and replacement site mutations occur, the rate of change for replacement sites has been five times that for silent sites. This result is extremely unlikely to have come about just from chance. Positive selection thus acted sometime in the history of this branch of the evolutionary tree. (The technical term for branches of the evolutionary tree is "lineages.") Other lineages of Antarctic fish also display the same pattern of a greater rate of replacement site than silent site substitutions in their hemoglobin, thus indicting the action of positive selection; the pattern, however, is most clear in the dragonfish lineage. One important caveat is that although this test can detect the action of positive selection, it cannot pinpoint the exact nature of positive selection by itself. What is clear is that there has been positive selection operating on features of the hemoglobin.

■ The Best Defense Is a Good Offense

Biologists have found this signature of the operation of positive selection in a variety of molecules among very diverse forms of life. Let's examine a few other molecules that show this same hallmark. A good place to start is with biological arms races. For example, most plants are engaged in a never-ending struggle against predators, parasites, and pathogens. The plants evolve adaptations to defend against or allow improved toleration of their natural enemies. In turn, the natural enemies evolve counter-measures to evade the plant defenses. Such arms races can last for millions of years, going through numerous escalating cycles of attack and resistance. Genes that are involved in plant resistance would thus be good candidates to look for the footprints of positive selection.

Fungi are one particular group of plant natural enemies that can be quite damaging; one famous example, *Phytophthora infestans*, was responsible for the Great Potato Famine in Ireland during the middle decades of the nineteenth century. To defend against fungi, some plants directly attack the hard substance used in the cell walls of fungi, known as chitin. Plants can degrade chitin by use of specialized enzymes known as chitinases; these enzymes destroy chitin without affecting plant cell walls, which generally contain cellulose rather than chitin.

Tom Mitchell-Olds and other researchers at the Max Planck Institute for Chemical Ecology recently examined chitinases from several species of plants in the mustard family, including *Arabidopsis thaliana*.[3] This plant species is sometimes called "the green *Drosophila*" because geneticists have studied it more extensively than they have any other flowering plant. The researchers found many instances of more replacement changes than silent ones occurring between two species. In eleven pair-wise comparisons, this excess of replacement changes was tested and found to be statistically significant—that is, unlikely to be due to chance. In all of these statistically significant cases, replacement sites changed three or more times faster than did silent sites. This is strong evidence for positive selection driving many adaptive changes.

The Max Planck researchers were able to go further with the study by the use of more sophisticated models that enabled them to pinpoint the specific amino acids that were most likely to be the targets of positive selection. They also had access to X-ray crystallography pictures of the protein. Although proteins are strings of amino acids, their actual structure is far more complicated than a linear chain. Proteins fold into convoluted three-dimensional structures determined in part by their amino acid sequence, but also by the environmental conditions these proteins find themselves in; physical chemists are just now learning the "rules" that govern protein folding. X-ray crystallography has been and remains a powerful tool to examine the structures of complex molecules, including proteins. The Max Planck researchers found that most of the adaptive replacements in the chitinases were occurring in the active site cleft, an unusual pattern; in most other instances where positive selection acted on a protein, the adaptive changes occurred far away from the active site. The authors suggest that positive selection may be driving changes in the active site because the fungi attacked by the chitinases defend themselves via forms of chemical resistance, such as trying to inhibit the chitinase.

■ The McDonald-Kreitman Test and Flies That Can Hold Their Liquor

A significant excess of replacement site changes over silent site changes is a very clear signature of the action of positive selection. Yet the absence of this excess of replacement changes does not necessarily mean that positive selection has not been acting. Suppose that both positive and negative selection had been acting on replacement mutations in a molecule. This negative selection would act more strongly on the replacement than it would on silent mutations. By acting on the replacement mutations, negative selection would reduce the replacement substitution rate. Like rain washing away footprints, negative selection would obscure the signature of positive selection (the excess of replacement substitutions).

Good detectives are sometimes able to follow the trail of a suspect even after rain would have washed away any footprints; they do this by looking at other signs of human movement, such as the disturbance of leaf clutter. Similarly, molecular evolutionists have been able to use alternative means to detect positive selection even when negative selection and other evolutionary forces have obscured the clearest sign of positive selection (excess replacement over silent site changes). During the 1990s, several tests were proposed to detect the action of positive selection even in the absence of an excess of replacement substitutions. One of the earliest and easiest of such tests is the McDonald-Kreitman test, named after the scientists who developed and applied it.

While still a graduate student at Harvard during the early 1980s, Marty Kreitman became renowned in evolutionary genetic circles by producing the first detailed study of variation at the DNA level within a species. Kreitman's choice of the fruit fly *Drosophila melanogaster* as his study organism should come as no surprise; geneticists have devoted decades of study to that fly. The gene that Kreitman studied was also very well known—the *Adh* gene that codes for the enzyme alcohol dehydrogenase, which breaks down ethanol (the alcohol in alcoholic beverages) into acetaldehyde. *Drosophila melanogaster*, like many fruit flies, develops as larvae inside rotting fruit, where they feed on the yeast growing on the surfaces of fruits. Even as adults, these flies roam around the fruit eating yeast, mating, and in the case of females, depositing eggs. The yeast convert the sugars in fruit into ethanol and other chemicals. Thus, if the flies (and especially the fly larvae) lack adequate detoxifying enzymes such as alcohol dehydrogenase, they will become too drunk to eat or move properly.

Kreitman, in his 1983 study, found substantial variation in nucleotides at the *Adh* gene.[4] This study by itself did not deal with the adaptive value of the variation. The value of Kreitman's study was that it paved the way for thousands of other subsequent studies that have examined variation within species at the level of DNA.

Fast forward to the early 1990s. Marty Kreitman was at his first faculty position at Princeton University and was still engaged in investigating the patterns of DNA variants in *D. melanogaster* and its closest relatives at *Adh* and other genes. He took on as a postdoctoral fellow John H. McDonald, who had studied marine organisms for his doctoral work at the State University of New York at Stony Brook. McDonald and Kreitman strongly suspected that positive selection had been working on *Adh* in part because of *Drosophila melanogaster* had a higher tolerance for ethanol than did *Drosophila simulans*.

In 1991, McDonald and Kreitman proposed a test for positive selection based on examining both divergence and polymorphism—variation among and within species.[5] Recall that changes in the mutation rate change divergence among species and polymorphism within species in the exactly same way. Double the mutation rate and both polymorphism and divergence double. The extent of negative selection (functional constraint) also has identical effects on divergence and polymorphism. A gene that is under high functional constraint will have reduced polymorphism as well as divergence, and to the same extent in each case. Positive selection, however, would increase divergence but would not increase polymorphism. In the McDonald-Kreitman test, silent sites are used as a control. The ratio of polymorphism (within species) to divergence (between species) for replacement sites is compared with that same ratio for silent sites.

McDonald and Kreitman compared polymorphism within *Drosophila melanogaster* to divergence between *Drosophila melanogaster* and two of its close relatives, *Drosophila simulans* and *Drosophila yakuba*, for both silent and replacement sites. They found a striking pattern; at silent sites, polymorphisms within *melanogaster* outnumber fixed differences between species by 42 to 17, respectively. But when they looked at replacement sites, they saw a different pattern: of the nine variable replacement sites, only two were polymorphisms within species and the other seven were fixed between species. More than eight times as many differences at fixed replacement sites between the species occurred than what had been predicted by the ratios seen at the silent sites. This discrepancy is highly unlikely to have occurred by chance.

One the one hand, a faster rate of evolution of replacement sites to silent sites is a very clear signal of positive selection. Such a pattern would be very difficult to explain by any process other than positive selection. On the other hand, the signal from the McDonald-Kreitman test of a significantly greater ratio of divergence (fixed differences) to polymorphism in replacement sites than in silent sites could conceivably be due to processes other than positive selection. In fact, McDonald and Kreitman proposed an alternative hypothesis for this pattern. What if, they supposed, the ancestral population sizes of both species were much smaller than they are currently? They also supposed that many slightly deleterious replacement mutations had occurred. Recall that the random force of genetic drift is stronger when population sizes are low. Because the ancestral populations were very small, slightly deleterious mutations could become fixed in the populations via random genetic drift. Fixation of these slightly deleterious mutations would result in many replacement differences between the species. When the populations expanded, the slightly deleterious replacement mutations would be weeded out by negative selection. These would not appear as polymorphisms. If you think this scenario sounds complicated, you are correct. In fact, McDonald and Kreitman said so in justifying why they favored the adaptive explanation over the slightly deleterious model to explain their results.

> This slightly deleterious model requires so many assumptions about selection coefficients, population sizes, and times of population expansions, that we prefer the simpler explanation, the occasional fixation of an adaptive mutation.[6]

In other words, McDonald and Kreitman invoked Occam's razor to support their choice of explanations.

A virtue of the McDonald-Kreitman test is that it is simple to perform—requiring nothing more than simple arithmetic, and can be calculated in a few minutes with a hand-held calculator.[7] The data required are just several sequences within one species and a comparison sequence from a closely related species.

■ How to Perform the McDonald-Kreitman Test

So, let's look at the actual data collected by John McDonald and Marty Kreitman as shown in the table below. The top row shows the numbers of amino-acid replacement changes in the *Adh* gene; the bottom shows the numbers of silent changes. The left-hand column displays the number of fixed changes between species, the right shows the number of polymorphisms within *Drosophila melanogaster*.

7	2
17	42

There are 42 silent polymorphisms within *Drosophila melanogaster* and 7 fixed replacement differences between species. In addition, the researchers observed 2 replacement polymorphisms and 17 silent fixed differences.

Based on the neutral theory, the ratios of the fixed differences to polymorphisms should be the same for the replacement differences ($7/2 = 3.5$) and the silent differences ($17/42 = $ just over 0.4). The ratios aren't the same, but are they statistically different?

To figure that out, we first need to determine the expected values for each of the four entries. The expected value for the number of replacement (amino acid changing) substitutions is equal to the total number of replacement changes ($7 + 2 = 9$) times the total number of fixed differences ($7 + 17 = 24$), divided by the grand total ($7 + 17 + 2 + 42 = 68$). The answer is 3.176, which we can round to 3.2.

We can do the same for all four cells of the table, and we get

3.2	5.8
20.8	38.2

■ **How to Perform the McDonald-Kreitman Test** *(continued)*

Based on McDonald and Kreitman's data, we would expect to observe 3.2 replacement substitutions, 5.8 replacement polymorphisms, 20.8 fixed silent changes, and 38.2 silent polymorphisms.

The final step is to compare the actual observed numbers (the top table) with the expected numbers (the bottom table) for all four cells.

$$+3.8 \qquad -3.8$$
$$-3.8 \qquad +3.8$$

Compared with the expected values, 3.8 more fixed replacement changes occurred. We see deviations of the same magnitude—but sometimes of different sign—for the other cells in the table; for instance, there are 3.8 fewer than expected replacement polymorphisms. Can these deviations be explained by chance, or are they manifestation of some other force operating? To answer that question, McDonald and Kreitman used a goodness-of-fit statistical test; these tests, as their name suggests, tell how well the expected fit the observed. The details of these tests are beyond the scope of this book, but according to the G test (a well-known goodness-of-fit test), the probability is less than 1% of having the observed values differ as much from the expected values as the data in the McDonald-Kreitman dataset did. Therefore, can be quite confident that the excess of amino-acid changing substitutions is not due solely to chance.

During the decade following publication of McDonald and Kreitman's article in *Nature*, numerous other tests for positive selection were proposed: the Fay and Wu's test, the Fu and Li's test, and the Tajima's D, named after their inventors. Most of these tests are more complicated than the McDonald-Kreitman, but the central principle is the same: testing for positive selection by looking for specific deviations from the expectations of Kimura's neutral theory. The results of such tests will appear several times in latter chapters and demonstrate that detecting the action of positive selection can shed light on the natural history of humans and other biological organisms.

■ **Selective Sweeps and the Hitchhiker's Guide to the Genome**

Before we discuss global patterns of selection that act on the genome, let's consider what happens when a new and advantageous mutation arises in the population. Most of the time that new mutation—despite being advantageous—will be lost from the population because new mutations start off as just a single copy and are easily lost. For instance, there may be a mutation that enables a fish living in cold water to be better able to tolerate cold, but if the individual that harbors that new mutation is eaten before it gets a chance to breed, that's the end of that mutation. Advantageous mutations are much more likely to become fixed in the population than are neutral ones, which in turn, have a far better probability of fixation than do even slightly deleterious ones. Evolution is often a game of probability, but as every poker player knows, some events (getting two pair) are more probable than others (getting a royal flush).

So let's consider that the new, beneficial mutation has survived the initial stages when it was present in only a few individuals. From this point on, its fate is sealed; it will rapidly increase in frequency in the population until it is fixed, assuming that the environment stays consistent. How rapidly this increase in frequency of the favored variant occurs, what geneticists some-times call a "selective sweep," depends mainly on the strength (and direction) of selection. Suppose two alternative genetic variants (A and a) are possible. Further suppose that 50% of the individuals that had two copies of A (geno-type AA) survived to adulthood. Also suppose that 45% of those individuals with one copy of A and one copy of a (genotype Aa) survived, and only 40% of those with two copies of a (genotype aa) survived. If no differences exist among these genotypes with respect to reproduction or mating ability, the relative fitness of genotype AA would be 1, that of Aa would be 0.9, and that of aa would be 0.8. Under these conditions, the frequency of the A variant can go from 0.01 to 0.99 in under 100 generations. If the selective differences among the genotypes are smaller, the change in frequency will be slower; but even when the differences are a couple tenths of a percent the advantageous variant will be able to sweep through the population in only a few thousand generations.

An important consequence of the relative rapidity of these selective sweeps is that other genetic variants close on the chromosome to the advantageous genetic variant will also rise in frequency, and could also become fixed in the population. These other variants would not necessarily be of any fitness ben-efit to the organism; indeed, these variants hitchhiking along with the linked advantageous variant could well be deleterious. The rise in frequency of hitchhiking variants is just a happenstance of the other variant that they are near. Genetic recombination is the only thing that would prevent them from becoming fixed. Recombination occurs every generation, but the quicker the selective sweep, the less time recombination would have to prevent the fixation of hitchhiking variants.

This "hitchhiking" phenomenon and the importance of recombination were known decades before we had any data and before we could do DNA sequencing on a massive scale. Moreover, in a 1974 paper, the brilliant British evolutionary geneticist John Maynard Smith noted that through hitchhiking, the selective sweep would have consequences similar to that of a population bottleneck: it would reduce genetic variation.[8] Unlike the reduction in variation that arises from a population bottleneck, this reduction in variation associated with a selective sweep would be localized to the chromosomal region near the variant that swept through. How far this region extends depends upon the strengths of selection and the frequency of recombination.

Darwinian detectives thus can infer a recent selective sweep when they observe a region of much reduced genetic variability on the chromosome near regions with higher extents of variation. To be detected as such, the sweep must have taken place fairly recently because in every generation after the sweep, new variants will again accumulate due to mutation.

One chromosome of *Drosophila melanogaster* is much smaller than the others; genes on this chromosome, sometimes called the "dot" chromosome due to its small size, experience essentially no recombination. This lack of recombination means that the entire chromosome acts as an evolutionary unit; with respect to evolutionary transmission, it is a single "gene." Any variant that sweeps to fixation due to selection will carry with it all other associated variants. If selective sweeps happen relatively often, one would expect very low levels of genetic variation along the whole chromosome.

Consistent with the predictions, genes across the dot chromosome show remarkably low levels of variation. For instance, consider the dot-chromosome gene *cubitus interruptus* (*ci*), a gene involved in pattern formation in early development of flies. Very little or no genetic variation is found within any of one of the four species *Drosophila melanogaster*, *Drosophila simulans*, *Drosophila sechellia*, and *Drosophila mauritiana*.[9] This absence of variation cannot be explained by functional constraint, because normal levels of genetic divergence among the species are observed for this gene. Moreover, the lack of variation is seen in both amino-acid changing sites and those silent sites that don't affect amino acid sequences. Other dot-chromosome genes display the same pattern: zero to very low levels of polymorphism within species with normal to high levels of divergence among species. The best explanation for this phenomenon is the consequence of selective sweeps and genetic hitch-hiking. In cases such as the dot chromosome, in which a large chunk of DNA has little or no recombination, one cannot infer which of the genes was the target of selection.

■ Linkage Disequilibrium and Selection

While selection is in the process of driving a genetic variant to high frequency, the selected variant carries along other variants that are genetically linked with it. In this situation, one would expect that the variants at one site would be highly correlated with variants at other nearby sites. An hypothetical example of such correlations would be that if a chromosome sampled has nucleotide "A" at site 1, it would be more likely to have nucleotide "G" at site 2 (1,563 nucleotides away) than if it had a different variant at site 1. Correlations, in general, provide predictive power. For example, weather patterns between consecutive days are correlated, and these correlations give us some ability to predict the weather the next day. If it's colder than normal on Monday, chances are good that Tuesday will be cold as well. The same is true with these correlations among genetic variants: knowledge of what variant is present at one site in an individual provides predictive information about what variants are present at other nearby sites.

With respect to weather forecasts, predictive power falls off with distance. It's quite likely that the weather an hour from now will be more or less the same as the weather now. It's much harder to predict the weather a week from now on the basis of the weather today. The same is true in genetics; as we shall see, linkage disequilibrium (the genetic predictive power) tends to fall off with increasing genetic distance.

Population geneticists were aware of such correlations decades before DNA sequencing was possible. In fact, they had developed a theory that predicts how these correlations would arise and disappear. Specifically, they found that in the absence of evolutionary forces (selection, genetic drift, and mutation), genetic recombination should result in the erosion of correlation among variants at different sites. These population geneticists called such correlations "linkage disequilibrium" because the patterns were usually only observed when the variants were genetically linked (close on the same chromosome). At that time, little was done—little could be done—to investigate patterns of variation within genes.

The Human Genome Project and its spin-offs changed this. Linkage disequilibrium (known also as LD), once an arcane subject in population

```
        1           2                        3
AGGTAAGTAGCGATCCACGGTA...........TAGGGAAT
AGGTAAGTAGCGATCCACGGTA...........TAGGCAAT
AGGTAAGTAGCGATCCACGGTA...........TAGGGAAT
AGGTAAGTAGCGATCCACGGTA...........TAGGCAAT
AGGTAACTAGCGATCAACGGTA...........TAGGCAAT
AGGTAACTAGCGATCAACGGTA...........TAGGGAAT
AGGTAACTAGCGATCAACGGTA...........TAGGGAAT
AGGTAACTAGCGATCAACGGTA...........TAGGCAAT
```

Figure 4.2

Example of linkage disequilibrium. Eight different DNA sequences are collected from a population. Three sites are polymorphic and are shaded in gray. The other sites are monomorphic; no variation exists at those sites. The ellipses denote a chunk of missing DNA. The variants in site 1 and the variants in site 2 are in linkage disequilibrium with either site 1 or site 2; the variants present at site 1 and site 2 are not associated with G or C at site 3 (the frequencies of G and C at site 3 are the same regardless of whether site 1 has a G or a C).

genetics, is now of great interest to the human genetics community because these tight correlations among linked variants can be used to map genetic variants associated with traits, including medically important ones. In such mapping, researchers search for regions of the genome where affected individuals are unusually similar in the genetic variants they possess. The existence of LD is an aid in these mapping attempts because the presence of highly correlated variants provides more statistical power.

In principle, observed patterns of LD can be used to infer the action of positive selection. The presence of LD itself does not signify that positive selection has acted; even without positive selection, other evolutionary forces (particularly random genetic drift) can lead to LD.[10] Genetic drift, however, should have similar effects across the whole genome, whereas the effects of a selective sweep should be localized. In humans, LD due to genetic drift tends to disappear after a distance of several thousand nucleotides. Several exceptions of this pattern, however, have been found wherein LD can extend much further along chromosomes. The question is whether these exceptional cases reflect the variability inherent in a random process such as genetic drift or indicate the action of something other than drift (namely, positive selection).

Three different genes involved in the immune system of *Drosophila* all exhibit strong patterns of LD in a natural Californian population of *Drosophila simulans*.[11] In contrast, much less LD was found in an African population of this species. This pattern is consistent with positive selection that is currently selecting for a variant or variants within the California population, but not the African population. Other explanations for the high levels of LD in the population are possible, so we cannot definitively state that positive selection has acted on these genes based on the LD pattern alone. In conjunction with other evidence, this pattern can point to the operation of selection.

LD created by directional selection will only persist for as long as polymorphism exists in the population. Given that the eventual result of directional selection is the loss of genetic polymorphism, LD as a signal for directional selection has limited utility. Balancing selection—where selection maintains two or more genetic variants at a given site—can also create LD; and as we will see in the next chapter, LD created by balancing selection can persist indefinitely.

Regardless of its cause, the presence of LD creates a structure in the patterns of genetic diversity within a gene or gene region by limiting the number of combinations of variants found in populations. Consider a gene region with 20 polymorphic sites. If each of these variable sites has two different variants, over a million possible combinations are possible. In actual samples, almost regardless of the species studied, far fewer actual combinations of variants appear. Strong linkage results in a few groups of closely linked genetic variants that are inherited as a unit. These groups are called "haplotypes," short for the phrase "haploid genotype."[12] Each individual has two haplotypes (one from the chromosome they inherited from each parent).

■ Whither the Neutral Theory? (Kimura's Legacy Part 2)

We have explored several cases in which a strong signal has been found for the action of positive selection. How frequent is positive selection across the genome as a whole? Some recent studies suggest that positive selection may have been responsible for the fixation of a large proportion—possibly as much as a half—of amino acid changes. For instance, Nicolas Bierne and Adam Eyre-Walker from the University of Sussex estimate that 25% of the divergence in replacement (amino-acid-changing) sites between different *Drosophila* species was driven by positive selection.[13]

What does the frequent substitution of advantageous variants imply for the fate of the neutral theory and Kimura's legacy? Does this mean that the neutral theory is dead?

First, these global estimates of the relative proportion of advantageous substitutions rest upon a number of assumptions and should still be considered preliminary. In fact, the confidence interval for the estimate by Bierne and Eyre-Walker above is huge: the percentage of replacement substitutions that were driven by positive selection could be as high as 45% or as low as 5%. Although the higher number challenges at least some aspects of the neutral theory, the lower number is perfectly compatible with Kimura's view. But let's assume that most amino-acid substitutions are advantageous, as these studies suggest. That doesn't invalidate all of Kimura's neutral theory.

Even if 45% of replacement *substitutions* were driven by positive selection, this mode of selection is still much rarer than negative selection for most genes. Recall that the number of substitutions is far fewer than the total number of mutations, as only a very small percentage of mutations become fixed. It is still clear that most new mutations are either neutral or deleterious, with only a relatively small fraction being advantageous. For instance, as we will see in chapter 8, comparison of the human and chimp genome sequences shows a high level of functional constraint acting in primates. In the human and chimpanzee genomes as a whole, silent substitutions are about four times more common than those that alter the amino-acid sequence; because silent changes are likely to be near neutral, we can thus declare that negative selection dominates the evolution of protein-coding genes.

Not only are most new mutations either neutral or detrimental, but so are most genetic variants existing within populations. To a rough approximation, the predictions from the neutral theory regarding the relationship between genetic polymorphism and effective population size hold.

Still, the most important contribution of Kimura's neutral theory has been its effect on the science of molecular evolution. Although positive selection appears to be more common than Kimura may have thought, his model—one that assumed the rarity of positive selection—continues to be very fruitful. The formation of the neutral theory enabled researchers to establish numerous new research programs, including the generation of many tests that were and can be used to demonstrate positive selection.

5 ∎

Balancing Selection and Disease

[W]e cannot know the over-all importance of balancing
selection by demonstrating that it exists. Of course it exists.
The problem is, what proportion of the observed genetic
variation is maintained by selection.

—Richard Lewontin (Lewontin, 1974, p. 232)

∎ Sickle Cell Anemia: The Textbook Case for Balancing Selection

Each year, about a quarter of a million babies are born with sickle cell anemia.[1] The name of this devastating disease comes from the "sickling" of the red blood cells; instead of the normal round shape, most of these cells in a person with sickle cell anemia are crescent-shaped and stiff. The spleen rapidly destroys these sickle-shaped cells, and this destruction leads to anemia because the body cannot keep up the production of new red blood cells. In addition to the anemia, individuals with this disease have swollen hands and feet, episodes of pain, and frequent infections (due to spleen damage). Because of red blood cells sticking together on the walls of blood vessels, these individuals are at extremely high risk from heart attacks and strokes. In the developed world in recent years, with widespread vaccinations and prophy- lactic penicillin treatments, the survivorship of individuals with sickle cell anemia has dramatically increased; at end of the twentieth century in the United States, the life expectancies of afflicted individuals reached the forties. In regions without access to modern medicine, most people with sickle cell anemia do not survive past childhood.

Sickle cell anemia was among the first described "molecular" diseases. In 1949, using protein electrophoresis and other techniques, Linus Pauling and his coworkers showed differences in the electrical charge of hemoglobin from people with the anemia as compared with those of disease-free people. This finding allowed Pauling's group to develop a diagnostic test for the disease.[2]

Here's how Pauling's biographer, Thomas Hager, described the feat: "By pinpointing the source of a disease in the alteration of a specific molecule and firmly linking it to genetics, Pauling's group created a landmark in the history of both medicine and genetics."[3]

By 1957, Vernon Ingram had traced the defect to a single amino acid change in the beta chain of hemoglobin, and thus one nucleotide change in DNA. Individuals with sickle cell anemia have the amino acid valine instead of the normal amino acid glutamic acid in position 6 of the beta-hemoglobin.[4] The normal variant of hemoglobin is called hemoglobin A, and the sickle variant is called hemoglobin S. This small change in amino acid sequence is enough to cause a difference in the charge of the protein, which in turn leads to myriad effects as evidenced by the range of symptoms in sickle cell anemia. The phenomenon of multiple effects from a single genetic change, what geneticists refer to as pleiotropy, is quite common in genetics.

Hemoglobin S, the sickle-cell genetic variant, is recessive to the normal variant; individuals with one copy of hemoglobin A and one copy of hemoglobin S (that is, heterozygotes) do not acquire the disease. Most heterozygotes for hemoglobin S do not notice anything; those that do usually experience only mild effects, apparent most often when exercising vigorously or during other cases of lowered oxygen supply. The sickle genetic variant is found in about two million Americans, mostly of African descent. Such unaffected individuals with the variant are often called carriers. If two carriers mate, their children each will have a one-quarter probability of inheriting the disease. (Children of a carrier and a noncarrier have a 50% probability of being carriers, but none will acquire the disease.)

Hemoglobin S is found almost exclusively in people whose ancestry traces back to Africa, parts of the Middle East, or the Indian subcontinent; it is very rare elsewhere. Moreover, its frequency is higher in populations from Africa than among African Americans. One in 500 births in African Americans has the anemia, but the frequency of people with the trait (heterozygotes) is around 10%.

Why should this genetic variant that is obviously severely deleterious in homozygotes be able to persist at such high frequencies? Some diseases are explained by a balance between recurrent mutation increasing the frequency of the disease variant and negative selection weeding them out. Such diseases, however, are typically much rarer than sickle cell anemia; mutation rates usually aren't high enough to balance out so much negative selection. Moreover, why should sickle cell anemia be so limited geographically?

It turns out that the answer to both questions is malaria.[5] This disease infects over 300 million people worldwide, and more than a million die from it each year. Although malaria was not uncommon in Europe and the Americas as recently as the nineteenth century, it is most prevalent today in Africa, the southern part of the Arabian Peninsula, India, and the Mediterranean. Hemoglobin S is found only in peoples with ancestry from regions where malaria is prevalent. For instance, this variant is much rarer in peoples originally from the highlands of Africa, where malaria is rare or absent, than it is in the lowland populations. Malarial infection occurs upon bite of an *Anopheles* mosquito harboring the parasite *Plasmodium*, a single-celled organism with a complex life cycle. Once within a human host, this parasite will infect a red blood cell, and reproduce asexually inside the cell. The offspring will cause the infected cell to burst open, allowing the offspring parasites to be released and infect more red blood cells.

Heterozygotes for the sickle variant have increased protection from malaria. They are not completely immune from the disease, but the disease is far less deadly to them. Although the exact mechanism(s) for how heterozygotes are less susceptible to malaria is still not fully known, we have some clues as to how their protection arises. First, the *Plasmodium* parasites grow and develop at a slower rate in heterozygotes as compared to nonsickling homozygotes. Moreover, when the red blood cells of heterozygotes are infected with *Plasmodium*, they tend to sickle and are thus usually quickly destroyed by the spleen. Recent work suggests that the malaria protection in heterozygotes also may involve the immune system.[6]

Thus, this genetic polymorphism is maintained in areas of the world where malaria is prevalent because the heterozygote has a higher fitness than either homozygote. The homozygotes for the sickle variant die young from sickle cell anemia, and the homozygotes for the normal variant are at increased risk of dying from malaria. The heterozygotes have the best of both worlds.

A striking feature of the sickle genetic variant as an anti-malarial defense is just how wasteful it is. Would an Intelligent Designer design such a mechanism, one that would cause multitudes of painful deaths each generation? One might argue that perhaps hemoglobin S is the only genetic defense to malaria that could arise. But that is just not so—some alterations in the hemoglobin afford protection against malaria without causing sickle cell anemia. In fact, hemoglobin C provides protection against malaria without causing anemia and, like hemoglobin S, is the result of a change in the same position 6 of the beta hemoglobin chain. (Hemoglobin A has glutamic acid, hemoglobin C has lysine, and hemoglobin S has alanine).[7] Hemoglobin C reaches frequencies over 25% in some areas of West Africa, far exceeding the frequency of sickle cell anemia. So why did hemoglobin S, with its deleterious effects in homozygotes, spread throughout a large section of the world?

The answer is that when the sickle genetic variant first appeared, its frequency was very low. The variant thus would have been present almost exclusively in heterozygotes and not in homozygotes. (The proportion of homozygotes of a variant in a population is roughly equal to the square of the variant's frequency, and a very small number times a very small number is a miniscule number.) Positive, not balancing, selection would have increased the frequency of the sickle genetic variant until an appreciable number of homozygotes for the sickle variant were in the population. The frequency of hemoglobin S then continued to climb until the deaths due to the sickle cell anemia in hemoglobin S homozygotes balanced out the lives saved by the malaria protection in heterozygotes. Because (even in areas where malaria is very common) the homozygotes for hemoglobin A have a higher fitness than the homozygotes for hemoglobin S, hemoglobin A is much more common than hemoglobin S at equilibrium.

Where malaria is not prevalent, hemoglobin S does not provide a fitness benefit to heterozygotes. Because it causes sickle cell anemia in homozygotes, one should expect the frequency of the hemoglobin S variant to decline in regions without malaria. Population genetics theory does predict that, all things being equal, the rate of decline in frequency of a recessive deleterious genetic variant should be slower than that of a dominant deleterious variant. Nonetheless, there should be a decline. And if you look at the frequency of the sickle cell genetic variant in African Americans, it is lower than that of Africans. Part of that difference is because many African Americans have mixed ancestry, but part of it reflects the negative selection that has acted on the hemoglobin S variant in an environment where malaria is not prevalent.

Sickle cell anemia has become the textbook case for heterozygote superiority and for balancing selection in general. As discussed in chapter 3, many biologists between approximately 1950 and 1980 saw balancing selection as a major explanation for the maintenance of genetic variation. The problem for these advocates was that there was a whole lot of genetic variation, but precious few well-documented cases of balancing selection maintaining genetic variants. In fact, for a long time, sickle cell anemia was the only documented case! In 1974, Richard Lewontin, one of the developers of protein electrophoresis to study genetic variation, likened sickle-cell anemia to a "tired old Bucephalus" (Alexander the Great's horse) that is always trotted out in support of balancing selection. Lewontin went on to say: "Anyone who has taught genetics for a number of years is tired of sickle-cell anemia and embarrassed by the fact that it is the only authenticated case of overdominance available."[8] In other words, if balancing selection is so important, where are all of the other cases?

One such case appears to be associated with cystic fibrosis. Like sickle-cell anemia, cystic fibrosis is a recessive disease; just like in sickle-cell anemia, heterozygotes for the genetic variant that causes cystic fibrosis appear to have a selective advantage. In cystic fibrosis, mucus builds up in the lungs and intestinal tract. Prior to modern medicine, individuals with cystic fibrosis died in early childhood. One in 2,500 babies of European descent is born with cystic fibrosis, but it is much rarer elsewhere. Roughly 4% of people of European descent are heterozygotes, and thus carriers for cystic fibrosis, but have no or mild disease symptoms. The gene for cystic fibrosis, which was among the first human disease genes to be sequenced, codes for a protein called the cystic fibrosis transmembrane conductance regulator, CFTR for short.[9] This protein transports chloride ions across membranes. Its normal function ensures proper water balance. This function led to speculation that heterozygotes for the cystic fibrosis variant are better protected from dehydration that results from the diarrhea associated with cholera than those individuals homozygous for the "normal" variant. This hypothesis, however intriguing, currently is not as well supported as the sickle-cell anemia case study.

Just as evolutionary geneticists can infer positive selection from examining patterns of variation in DNA from populations, these researchers can infer balancing selection using similar methods that also rely on observing deviations from the expectations of Kimura's neutral theory. In the remainder of this chapter, we will discuss how these Darwinian detectives infer balancing selection from DNA sequence data.

■ HIV and the Chemokine Receptor

A recent study by Michael Bamshad and his colleagues on the regulatory region of the chemokine receptor is an exemplar of how multiple lines of evidence converge to support the action of one mode of selection in the region of the genome. But before we get to their study, we need to discuss some features of the immune system and the human immunodeficiency virus (HIV), the virus that causes AIDS.

Upon infection, abrasion, bruise, or any other challenge to the immune system, cells known as helper T-cells secrete chemokines, molecules that attract white blood cells to the sites of infection. Spanning the cell membranes of white blood cells, various proteins act as receptors for the chemokines. Although these chemokine receptor proteins play an important role in immune responses, viruses and other entities can subvert the receptors. And that's what HIV does; it initially gets into white blood cells via a particular chemokine receptor, known as CCR5.

In the middle 1990s, several research teams showed a link between variation in the gene that codes for CCR5 and HIV susceptibility; HIV susceptibility correlated with how much of the CCR5 protein was expressed on the surface of white blood cells.[10] Probably the most notable genetic variant of *CCR5* is one that has a deletion of 32 nucleotides, and hence is called delta-32. The protein arising from the deletion isn't just short a few amino acids; it is truncated and completely nonfunctional. Remember the triplet (3 to 1) code of translating nucleic acid information into amino acid (protein) information. Because 32 cannot be evenly divided by 3, the deletion is out of phase, and thus no functional product ensues. Hence, homozygotes for the deletion lack CCR5 molecules on the surface of their white blood cells. There are other chemokine receptors besides CCR5, so these white blood cells can respond to chemokines, and apparently the lack of CCR5 does not have detectable deleterious effects. When infected with HIV, individuals that are heterozygous for delta-32 proceed to full-blown AIDS at a much slower rate than normal; homozygotes for the deletion are virtually immune to HIV infection.

Intriguingly, delta-32 has a clinal pattern of distribution; its frequency is over 15% in some populations that are originally from Northern Europe, and is lower in Southern European populations, rare in the Middle East, and virtually absent in Africa. It seems surprising that a genetic variant that provides resistance to HIV would be least common in Africa, where HIV probably originated and is most prevalent currently. The high frequency of delta-32 in Europe, where HIV was not found until the late twentieth century, also appears odd. Certainly, the advantage this variant has with respect to resistance to HIV infection cannot explain its current geographical distribution. Moreover, based on the patterns of variation at nucleotides in the vicinity of delta-32, delta-32 appears to have arisen about 700 years ago (give or take a couple centuries), far older than the estimate of HIV's age. Something else must be going on.

Some speculate that the increase in frequency in delta-32 seen in northern Europe was because the deletion protected against the Black Plague of the fourteenth century. This plague hypothesis, although very intriguing, is still more speculative than proven; it is not quite clear how the deletion would give protection against the plague. The age of deletion delta-32 is consistent with a major plague epidemic, but it should be noted that the margin of error associated with this age is quite large. Its increase in frequency could have occurred 500 years ago; it could have occurred 1,000 years ago. It is quite clear, however, that delta-32 rose in frequency not because of its current protective effect against HIV but for some other reason.

Michael Bamshad and his colleagues had obtained DNA samples from several human populations (Africa, Europe, South Indian, and Asian, as well as a multi-ethnic sampling from the United States). They were particularly interested in a region of DNA that is nearby (approximately 1,000–2,000 nucleotides away) the *CCR5* gene but does not code for proteins. This region, however, is involved in the regulation of the expression of the *CCR5* gene itself; that is, regulation of how much protein is made from the gene, when, and in which tissues. In the 1990s, evolutionary geneticists started to pay increasing attention to regulatory regions near coding genes.

Consistent with the hypothesis of balancing selection, the *CCR5* regulatory region has many more single nucleotide polymorphisms (SNPs) than most regions of the human genome.[11] This pattern would be expected under balancing selection because this type of selection maintains genetic variants. Yet several other reasons could explain the increased levels of variation, apart from balancing selection. Therefore, in order to claim that balancing selection is occurring, one would have to dig deeper, as Bamshad and his colleagues have done. From further analysis of the patterns of variation in this sequence, they can point to several lines of evidence supporting the hypothesis that balancing selection is or has been recently operating on this regulatory region of the *CCR5* gene.

According to the neutral theory of molecular evolution, every gene in the genome should have the same ratio of polymorphism *within* species to the extent of divergence *between* species. One particular gene may have a higher mutation rate, and thus would be expected to have a high degree of polymorphism within species; that increased mutation rate, however, also would cause it to have a higher rate of divergence between species compared with other genes. But consider a gene that is under a great deal of negative selection and functional constraint. It wouldn't be variable within species, but it also wouldn't diverge much between species. In all cases, the ratio of polymorphism to divergence should be the same, except if the assumptions of the neutral theory are being violated.

If the regulatory region of *CCR5* were under strong balancing selection, the ratio of within-species polymorphism to between-species divergence in this region should be statistically greater than that same ratio for a different gene that doesn't experience balancing selection. Bamshad and his colleagues looked at these ratios in this regulatory region of *CCR5* and a noncoding sequence of another gene (*CYP1A2*), a gene they had no reason to suspect was under strong balancing or positive selection. In addition to the human samples, Bamshad and his colleagues had collected samples from a chimpanzee and a gorilla. These samples enabled them to get two different estimates of divergence (human to chimp, and human to gorilla). Using a statistical test, these researchers were able to show that the ratio of polymorphism to divergence was significantly greater in *CCR5*'s regulatory region than in *CYP1A2*, precisely what would be expected if the regulatory sequences were under balancing selection. Although this test by itself only indicates that the two genes have statistically significantly different polymorphism-to-divergence ratios, there is no other indication that the *CYP1A2* sequence deviates from neutrality. Furthermore, *CCR5* also shows similar statistically significant deviations when tested against other genes. These results are further evidence for some form of balancing selection acting on the regulatory region of *CCR5*.

Another clue that *CCR5*'s regulatory region was under balancing selection came from Bamshad's research on the patterns of genetic differentiation in this genetic region among and within different ethnic groups. Before discussing how such studies can provide evidence for balancing selection, let's back up and consider the partitioning of genetic variation within and among different populations and ethnic groups of humans. Three decades of studies of genetic variation—including human blood groups, protein electrophoresis patterns, and DNA—have all led to the conclusion that the vast majority of genetic variation (85–90%) among humans is *within* rather than among populations.[12] Only about 6% of the variation exists among the so-called major "racial" groups, and another 5–10% occurs among different populations within those "racial" groups. For this reason, most (but not all) evolutionary biologists consider race to be a social rather than a biological construct; this is why I placed the word "racial" within quotes. Our species exhibits a modest degree of genetic differentiation; although some differences among populations do exist, these differences generally consist of changes in the frequencies of genetic variants across populations, rather than populations possessing fixed differences.

Even the modest population differentiation found in most genes is not present in the chemokine receptor regulatory region; the frequencies of the different genetic variants are extraordinarily consistent among human populations in Africa, Asia, South India, and Europe. These results are what would be expected if balancing selection were operating on the *CCR5* regulatory region; balancing selection would act to make the frequencies of variants more similar among different populations, resulting in markedly reduced differentiation. Could it be this lack of differentiation is a fluke of the sampling? To control for this possibility, the researchers also looked at the extent of differentiation of 100 *Alu* transposable elements that are more or less randomly scattered throughout the genome from the same individuals. Because *Alu* elements are unlikely to have major fitness consequences, this control allowed the researchers to determine whether the lack of population differentiation is actually from the gene study or has something to do with the individuals sampled. They found the population differentiation for the *Alu* elements to be significantly higher than that of the *CCR5* regulatory region and typical for that of most genes. Thus, the lack of population differentiation at the *CCR5* regulatory region is a result of something happening at that site or a site that is close by.

The *CCR5* regulatory region also exhibits deviations from the neutral theory with respect to the frequency spectra of variants. Given a known amount of polymorphism in a genetic region, the neutral theory makes predictions about how many variants will be rare, of intermediate frequency, and common. In all of the Old World populations examined of the regulatory region, more genetic variants are present at an intermediate frequency than would be expected under the neutral theory. In the Asian, European, and South Indian populations, this deviation was statistically significant, meaning that it was very unlikely to be due solely to chance; the same trend was present in the African population, but the difference was not statistically significant.

Population subdivision could explain the pattern of an excess of variants with intermediate frequency, but that scenario is very unlikely; we know that the *CCR5* regulatory region shows virtually zero population subdivision. A more likely explanation is that the pattern is the result of some form of balancing selection.

Evolutionary geneticists can also obtain clues about the forces operating on genes by constructing networks of the different haplotypes of individuals. Suppose two haplotypes are identical to each other, except one has the nucleotide A at position 765 and the other has a G at that position. These two haplotypes could be exchanged for each other by just one mutational step, and thus would be neighbors in a haplotype network. A pair of haplotypes that differed at two sites would require two mutational steps to be exchanged with each other. Pairs of haplotypes that required three, four, and five mutational steps in order to be exchanged for the other would be progressively more distant neighbors. Haplotype maps could then be constructed by linking haplotypes such that all were connected by the shortest possible route.

In their analysis of the *CCR5* regulatory region, Bamshad and colleagues found two large clusters of haplotypes, separated by a minimum of seven mutational steps—quite a large distance. This extensive separation between the two clusters, what the authors call "deep genealogical structure," is not only further evidence that balancing selection has been maintaining genetic variation in the regulatory region of *CCR5*, but also that it has been maintaining that variation for a very long time. The two clusters probably diverged 2.1 million years ago. Although estimates of divergence time based on such studies do have wide margins of error, it is still safe to say that balancing selection has been operating in this genetic region for at least hundreds of thousands of years. The long duration of the operation of balancing selection suggests that this regulatory region may play a more general role in disease resistance.

What is the nature of the balancing selection occurring in *CCR5*'s regulatory region? Is it selection for rare genetic variants? Is it superiority of heterozygotes? Having one copy from each of the two clusters seems to afford some protection from the devastating effects of HIV, as such heterozygotes develop full-blown AIDS slower after HIV infection than do individuals who have both copies from the same cluster. But that result doesn't speak directly to the role this region played before HIV came on the scene. At this point, we do not know the details of the historical action of balancing selection.

■ **Other Cases of Balancing Selection**

Balancing selection also appears to have been operating for millions of years on the major histocompatibility complex (MHC), which is an essential component of the mammalian immune system. In humans, MHC is a complex of 120 genes that spans four million nucleotides on chromosome 6 and is known as the human leukocyte antigen system (HLA). (When speaking about this gene complex in humans, I will interchange the terms MHC and HLA.) Several of the HMC genes encode proteins that serve as cell-surface antigens, and these will bind short fragments of proteins from viruses. This enables the T-cells that we met earlier to recognize foreign entities.[13]

Genetic variants at this complex have been associated with a variety of diseases, including insulin-dependent diabetes, narcolepsy, and rheumatoid arthritis. These associations aren't all-or-nothing cases, as we saw with sickle-cell anemia and cystic fibrosis. Instead, the variants that a person has at a given HLA locus affect the susceptibility of acquiring one of those diseases.

Several signatures in the patterns of genetic variation implicate balancing selection operating on HLA. The first sign is the high to extraordinary high levels of genetic polymorphism observed at genes in the HLA complex. Some HLA genetic loci have hundreds of genetic variants! In addition, the HLA variants also have a frequency distribution that is more "even" than would be expected under the neutral theory; that is, too many variants with intermediate frequencies. Strong linkage disequilibrium also persists for considerable distances in this complex.

Not only does convincing evidence exist that HLA variants are actively maintained by selection, but there is also strong support for the notion that this balancing selection has been operating at HLA for a long time. Bolstering this claim is the extent to which haplotypes at HLA genes differ considerably from one another; for instance, the average difference between haplotypes at the *HLA-A* locus is 35 nucleotides.[14] This finding is a strong signal that balancing selection has been maintaining genetic variants for a very long time. How long? Well, some human HLA genetic variants are more similar to some chimpanzee variants than they are to other human variants. Balancing selection most probably has been going on since before chimpanzees and humans diverged from our common ancestor six million years ago!

What is the nature of the balancing selection that has been operating on HLA/MHC? This is less clear than the existence of balancing selection, but there is support for reproductive aspects of biology being agents of the balancing selection. In humans, the risk of spontaneous miscarriages increases as progressively more genetic variants are shared between the mother and the father. In addition, MHC appears to be involved in mate choice in mice. Some evidence suggests a similar role in humans, but that link is not as well established as it is in mice. Balancing selection on HLA also appears to be pathogen driven, but the casual mechanisms are less well known. Perhaps having two variants at a particular HLA locus would increase the range of pathogens that one's immune system could detect. Plausible arguments can be made for the proposition that possessing rare variants would be advantageous. The problem is that heterozygote advantage and rare variant advantage make exactly the same predictions with respect to how they would affect patterns of genetic variation.

Balancing selection is certainly much more rare than negative selection and is also likely to be considerably more rare than positive selection. Yet balancing selection does appear to be more common than once thought. By examining patterns of DNA data with various statistical tools, evolutionary geneticists are able to implicate (with varying degrees of certitude) that balancing is or has acted in quite a few examples. Although these cases are not limited to disease in humans, such cases appear to be disproportionately abundant. The large number of human cases probably reflects the intensity of focus in human evolutionary genetics. One possibility is that the high proportion of cases of putative balancing selection being associated with disease reflects the fact that human geneticists are particularly interested in disease.

Nevertheless, there are biological reasons that we should expect balancing selection to be associated with disease. Parasites and pathogens, with their short generation times and high reproductive excess, have great opportunity for rapid evolution. Humans and other large animals would be under continual selective pressure just to keep up with these rapidly evolving pathogens and parasites. Being heterozygous for genetic variants would allow an individual to be able to respond to a wider array of pathogens and parasites. As we saw in sickle-cell anemia and perhaps in cystic fibrosis, however, adaptations to the natural enemy may involve responses that work well in one dose (in heterozygotes) but have deleterious consequences in two doses (in homozygotes). Natural selection does not "think" ahead; when such variants are rare, those latter negative consequences in homozygotes would not matter. Finally, mathematical theoretical studies examining the evolution of resistance to parasites and pathogens suggest that such evolution may often involve balancing selection of the sort whereby rare genetic variants are advantageous. Although the story is far from settled, my hunch is that balancing selection more frequently operates in cases involving disease.

Human Origins and Evolution

Though he lived in the century before television and blockbuster movies, Darwin had mastered the art of the cliffhanger. His only remarks about human evolution in *The Origin of Species* were "much light will be thrown on the origin of man and his history." While Victorian society waited, Darwin took 12 years—almost as long as the gap between the two *Star Wars* trilogies—before writing *The Descent of Man and Selection in Relation to Sex*.[1]

Whereas Darwin was content to take his time and to be a quiet revolutionary, others were not. Known as Darwin's bulldog for his passionate defense of Darwin and evolution, Thomas Huxley was among the first to discuss human origins and evolution within the scope of Darwinian evolution. In his 1861 essay "On Relations of Man to the Lower Animals," Huxley spoke of the gap between humans and other animals: "It would be no less wrong than absurd to deny the existence of this chasm, but it is at least equally wrong and absurd to exaggerate its magnitude, and, resting on the admitted fact of its existence, refuse to inquire whether it is wide or narrow."[2]

When he finally came to discussing humans in the light of evolution, Darwin presented argument after argument (with varying degrees of success) that this chasm was much smaller than commonly believed. Darwin pointed to numerous cases of animals appearing to display traits that had been considered the province of human emotions, including sympathy toward others, loyalty, and an appreciation for beauty and music. Darwin's point was that our differences—though they may be great—were ones of degree, and not of kind.

What can genetics tell us about our species and the differences between us humans and our closest relatives? The chapters that follow in this section will focus mainly on what we can learn about our origin and history from studies of patterns of DNA sequences, but first some words on the fossil record are in order.

Thanks to a century and a half of work by dedicated paleontologists, we can point to an almost seamless fossil record that begins very near the time of our common ancestor with chimpanzees and continues to just about the present day. Some of the fossils are so close to the chimp–human split that the paleontologists have difficulty determining whether the fossil is in fact on the human lineage. Lest a creationist twist the preceding words around, let me say clearly: the paleontologists' struggle with deciding whether a fossil is or is not on the human lineage reflects the progress they have made in coming so close to the actual point where proto-humans and proto-chimps split. It is a success, not a failing, of paleontology.

The fossil record very clearly demonstrates that our family tree is not like the ladder of popular myth, but rather like a bush. To say that a chimp-like organism begat *Australopithecus* begat *Homo erectus* begat Neanderthals begat us, grossly oversimplifies the true history of our lineage. Indeed, throughout most of this six million year history, often two or three (and sometimes more) "species" of ancient human lived at the same time. I've put the word species in quotes because one obviously could not test whether these contemporaneous forms could interbreed (the common standard of determining species status). Paleontologists continue to argue about what constitutes a species, both in general terms and with respect to our lineage. It is clear, however, that rather different organisms lived side by side during most of our history.

Of course, the bushiness of our family tree is no longer apparent; we are the sole surviving species of our genus, and we are separated by a rather considerable gulf from our closest relatives. A fascinating discovery published in late October 2004, however, shows that some bushiness remained in our lineage until very recently—far more recently than previously thought. In 2003, researchers discovered an almost complete, 18,000 year-old skeleton in a cave in Flores, a remote Indonesian island.[3] This skeleton not only was far smaller than modern humans (a little over three feet tall), but also had a brain size that was small even for its diminutive body. Parts of six other individuals, which were also very small and ranged in age from 13,000 to 94,000 years old, were also discovered, along with stone tools. Formally known as *Homo floresiensis*, this species has been dubbed "Hobbit Man" due to its similarity to J.R.R. Tolkien's creatures. Dwarfism is a common feature of mammals that dwell exclusively on a small island. The most notable case is the dwarf elephants that are considerably smaller than thoroughbred horses.

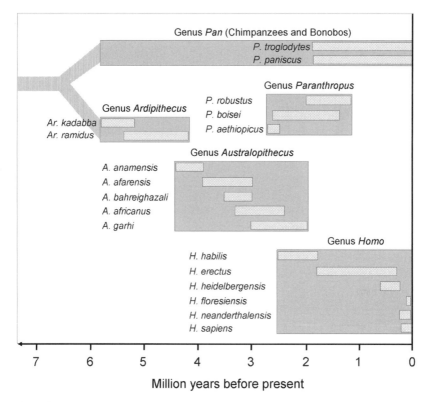

Figure H.1

The human "family tree." Dotted bars represent the approximate dates of the existence of species. Some disagreement exists among different authorities regarding the names and dates of some of the species (particularly within the more ancient species of the genus *Homo*). The evolutionary relationships among these fossil species are not specified. Note that *Homo neanderthalensis* and *Homo floresiensis* went extinct in the very recent past; if you look closely, you can see that their representative bars do not continue on to the present.

At 380 cubic centimeters (cc) in size, the brain of *Homo floresiensis* has less volume than a half-liter bottle of soda and is roughly the same size as that of chimps. Smaller organisms tend to have smaller brains, so an adjustment should be made for *Homo floresiensis'* small stature. When body size is taken into account, the Flores skull is comparable to that of *Homo erectus*. Most of the other characteristics of *Homo floresiensis*, including skull proportions and teeth shape, also support placement of this unusual skeleton within our own genus, *Homo*.

A common device used to capture the immense periods of time involved in evolution is to consider the age of earth to be a single year, and then plot events by such a calendar. Using that scale, the Cambrian explosion—the time when numerous forms of multicellular life made their first recorded appearance—occurred around Thanksgiving. The meteor that killed the dinosaurs 65 million years ago, marking the K/T boundary, occurred the day after Christmas, which the British celebrate as Boxing Day. By this scale, the branch of life (which biologists call a lineage) that led to us split from the branch that led to chimpanzees sometime shortly after noon on the last day of the year. The entire six million (give or take a million) year history of human separation from chimpanzees and the other apes is thus reduced to less than 12 hours!

But this scale soon becomes too grand for our purposes. By this reckoning, we poor (human) players don't even get a full second to strut and fret upon the stage. Let's be a little less ambitious in our use of scale; what if we were to collapse just those six million years since humans split from the chimpanzees into a single year, with January 1 as the date the lineages separated and December 31 as the present time? Here an 80-year lifespan becomes seven minutes, about half of what Andy Warhol promised us. We'll return to this device, the "human calendar," many times throughout this section.

Bipedalism, the habitual walking on two legs, evolved soon after we split from chimpanzees. Fossils that are over four million years old (early spring on the "human calendar"), on or close to our lineage, bear convincing signs of being bipedal. In fact, some biologists have speculated that the common ancestor was bipedal, and chimps subsequently lost bipedalism.[4] The evidence supporting this claim is far from solid, but it is an intriguing idea. Why did our precursors become bipedal so early on? There is no shortage of opinion regarding this question. Actually, we seem to suffer from an embarrassment of hypotheses regarding the selective advantage of bipedalism. Some hypotheses for the adaptive significance of bipedalism include the freeing of the hands to carry food, minimizing exposure to the sun, increased height associated with bipedalism, and the display of male genitalia coupled with the concealment of female genitalia!

Regardless of why we evolved bipedalism, our brains were not much bigger than those of chimpanzees when we started walking on two legs. Over the six million years of separation from chimps, our brains quadrupled in size. If this rate had been constant, it would mean an increase of only 2.3% every hundred thousand years. Obviously, our brains did not increase at a constant rate; there were jumps and starts, stalls, and perhaps even reversals. Nevertheless, when observed at a coarse scale, the rate of change of our brains appears almost linear. At no particular period of time in the history of the human lineage, can one say, "This is where our brains became human." Moreover, considerable variation exists within the brain sizes of living humans; some have brains with volumes comparable to a liter bottle of soda, while the brains of others are twice the size. Aside, from gross pathologies, little (at best) correlation exists between brain size and intelligence in modern humans (see chapter 9).

Even though our brain size continued to increase, our technology stalled about a million years ago. Jared Diamond has described this stagnation: "For most of the millions of years since our lineage diverged from that of apes, we remained little more than glorified chimpanzees in how we made our living."[5] Diamond proposed that we took a "Great Leap Forward" around fifty thousand years ago—a mere three days before the present on our human calendar. This was a time when sophisticated cave paintings, extensive use of wood tools, harpoons, bows and arrows, boats and trade routes, and beads and pendants were all added to our culture. All of these artifacts did not emerge at precisely the same time, nor did peoples of different geographical regions acquire the same level of culture simultaneously. Still, a dramatic change occurred within a few thousand years sometime around fifty thousand years ago.

We begin in chapter 6 with a discussion of how evolutionary geneticists traced our lineage by using a variety of genetic markers. We will focus on the mitochondrial DNA, which is transmitted only through the female line, and the Y chromosome, which is transmitted only through the male lines. We will see that the most recent common ancestor of all existing variants of this mitochondrial DNA—who must have been a woman—probably lived many thousands of years before the most recent common ancestor of Y chromosome types. Both of these individuals, and their progenitors, probably lived in Africa.

Most paleontologists think that the Neanderthals were a side branch of our tree; they view Neanderthals as our "aunts" and "uncles" rather than our "fathers" and "mothers." Much speculation has centered on whether the Neanderthal disappearance roughly 30,000 years ago was due to competition from, or even warfare with, our ancestors. Another contentious issue is whether Neanderthals passed along any DNA into our gene pool, and if they did, how much? These questions are addressed in chapter 7.

The discussion then turns to our closest living relatives, the common chimpanzee and the bonobo (previously known as the pygmy chimpanzees). We encounter the evidence showing that humans and chimpanzees are each other's closest relatives in chapter 8. This sets the stage for chapter 9, an account of recent studies that have detected positive selection upon the genetic changes along our lineage. What can these studies tell us about what has made us human?

We then return to comparatively recent history. Considerable debate exists as to when language evolved along the human lineage. For instance, Diamond has suggested that full-blown language may not have appeared until "The Great Leap Forward" 50,000 years ago. If this is so, perhaps we could trace back all languages to their common progenitor. The aims of chapter 10 are less ambitious. Here, we'll discuss how genetic and linguistic patterns do and do not correspond, focusing on peoples of Africa.

Although some species of ants will "farm" aphids in the production of a sweet substance called honeydew, and there are other examples of nonhuman agriculture, no other organism has used other animals and plants to the extent that we humans have. Moreover, in the process of using these animals, we also modified them considerably. Darwin, himself a pigeon fancier, made the analogy between the artificial selection we humans have imposed upon animals and plants and the natural selection imposed on organisms by natural forces. Evolutionary geneticists are increasingly able to determine the origins of domesticated animals and plants and to detect the signature of the artificial selection that we imposed on them through domestication. In chapter 11, we will see examples of such studies, focusing on the examples of dogs and maize.

■ **Recommended Reading**

There are several excellent general references on human evolution, which have been published within the past 20 years. A sample of these includes the following:

Dawkins, R. 2004. *The Ancestor's Tale: A Pilgrimage to the Dawn of Evolution.* Houghton Mifflin.

Diamond, J. 1992. *The Third Chimpanzee: The Evolution & Future of the Human Animal.* HarperCollins.

Diamond, J. 1997. *Guns, Germs, and Steel: The Fates of Human Societies.* W. W. Norton & Co.

Ehrlich, P. R. 2000. *Human Natures: Genes, Cultures, and the Human Prospect.* Penguin Books.

Marks, J. 2002. *What It Means to Be 98% Chimpanzee.* University of California Press.

Miller, G. 2000. *The Mating Mind: How Sexual Choice Shaped the Evolution of Human Nature.* Doubleday.

Relethford, J. H. 2003. *Reflections of Our Past: How Human History is Revealed in Our Genes.* Westview Press.

6 ■

Finding Our Roots
Did "Eve" Know "Adam"?

When you start about family, about lineage and ancestry, you
are talking about every person on earth.

—Alex Haley (in Marmom, 1977, p. 72)

■ **Roots**

Alex Haley traced back his ancestors to Africa, specifically Gambia, and to Kunta Kinte, his great-great-great-great-grandfather, who lived in the latter half of the eighteenth century. At age 17, Kinte was kidnapped, brought over to America, and sold into slavery. The recounting of his genealogical tale became the best-selling book *Roots*[1] in 1976 and, the following year, one of the most-watched television miniseries ever. The book and the miniseries (and its sequel) helped launch renewed interest in tracing family histories, especially among African Americans. Traditionally, genealogies were reconstructed by tracking census records, immigration records at places such as Ellis Island, slave-ship holdings, and so forth. In recent years, the Internet, both as a tool for communication and as a repository for data, has supplemented these traditional methods.

Yet people have always been interested in their family trees and history. Across virtually every culture, stories about ancestors abound. This passion, however, varies among cultures. In Iceland genealogy is practically a national obsession; the typical Icelander can trace her or his ancestry back several centuries further than the typical American. Kari Stefansson,[2] an Icelander geneticist who claims the tenth-century warrior and poet Egill Skallagrimsson as an ancestor, and his biotech company DeCode took advantage of such extensive genealogical records. With these records and DNA samples from a 100,000 Icelanders (more than a third of the entire population), Stefansson's company has been able to map genes to regions of chromosomes for several complex but common diseases including schizophrenia, Alzheimer's disease, Parkinson's disease, rheumatoid arthritis, and asthma.

DNA markers are yet another means to trace genealogies, and people increasingly use genetic markers to determine their origins.[3] Although these markers don't provide the rich detail of an individual life that can be gleaned from historical records, they can be very useful in tracing genealogies deep into ancient prehistory. In fact, for certain genes, geneticists now can determine where and when the common ancestor of all variants of that gene lived.

One small DNA molecule has been of great interest for those interested in tracing our genealogies—the mitochondrial DNA.

■ Mitochondria

Without the tiny organelles in our cells called mitochondria, we would be dead. These organelles are often called the "powerhouses of the cell" because they provide us with usable energy from the breakdown of sugars using oxygen. Absent mitochondria, our cells are unable to use oxygen to release energy from the breakdown of these sugars, and the breakdown of sugar in our cells in the absence of oxygen is many-fold less efficient. Several human diseases result from mutations in the DNA of the mitochondria.

The ancestors of mitochondria were once free-living organisms and share common ancestry with a subgroup of bacteria known as the alpha-proteobacteria. Somewhere between a billion and a half and two billion years ago, these mitochondria precursors formed an association with our distant ancestors, which were then just single-celled organisms. Either our ancestors engulfed the precursors of the mitochondria, or the ancestors to the mito-chondria invaded the larger cells. Either way, an association was formed. Most biologists think that the initial association was parasitic, and that it evolved from being parasitic, to benefiting both partners, to being obligatory. We cannot survive without the mitochondria and the energy they provide for us, and they cannot persist outside our cells. The mitochondria are dependent on us is partly because, in the course of this association, nearly all of what were the mitochondrial genes were transferred to the cell nucleus. In mammals, all that remains in the entire mitochondrial genome is just over 16,000 nucleotides—smaller than many individual human genes.

The manner in which mitochondria are transmitted probably would have given Brother Mendel fits. Instead of both parents participating, only the mother passes on her mitochondria. Males certainly have mitochondria; however, none of the male mitochondria, and hence, none of the paternal mitochondrial DNA, are transmitted when the sperm fertilizes the egg.[4]

Let's explore the implications of this mode of genetic transmission, what geneticists call maternal inheritance, with an example from my own family tree. My mother had three sons and passed on her mitochondria to us.

But because she did not have any daughters, her mitochondria will not persist; my brothers and I are dead-ends as far as her mitochondria are concerned. My mother received her mitochondria from her mother. My grandmother, in addition to my mother, had a son who had two children, one male and one female. Despite my maternal grandmother having had five grandchildren, her mitochondria will not be transmitted past this generation. My maternal grandmother's mother had a son and a daughter in addition to my grandmother; that daughter had only boys. Although my great-grandmother has a total of nine great grandchildren, her mitochondria reached dead-ends, and they will all be gone with the last of my generation.

For several historical reasons, molecular evolutionary geneticists interested in tracing genealogies within species and among closely related species focused first on mitochondrial DNA (mtDNA, for short). One reason is simply technical; sufficient quantities of mtDNA are easier to obtain than the same amounts of nuclear DNA. That's because each cell can contain hundreds of copies of mtDNA, but all cells have only a single copy of nuclear DNA. New technologies, developed in the 1980s and 1990s, ameliorated the necessity of obtaining large quantities of DNA in order to ascertain the sequence in question. One key technology is PCR, the polymerase chain reaction, made famous by the movie Jurassic Park and the O. J. Simpson trial.[5] This technique enables researchers to quickly mass-produce identical copies of DNA fragments up to a few thousand nucleotides in length. In principle, one could obtain billions of copies of DNA from a single DNA molecule in an afternoon with the use of PCR and associated technologies. Prior to the widespread use of PCR, however, the greater abundance of mtDNA over nuclear DNA made the former very attractive for evolutionary geneticists.

Another reason mitochondrial DNA was also the molecule of choice for studies of genealogies within species (or of relationships among closely related species) was because it tends to evolve much faster than most nuclear DNA. The faster rate of evolution for mtDNA is owing to the higher mutation rate of mtDNA that arises because the mitochondria lack the proofreading mechanisms to limit mutation present in the nucleus. Starting in the 1990s, evolutionary biologists interested in within-species genealogies turned increasingly to the use of microsatellites; these are stuttering segments of DNA that comprise many repeats of a short sequence of nucleotides (such as GATTC, repeated numerous times in succession: GATTCGATTCGATTCGATTC...). Microsatellites have high mutation rates (higher than that of mtDNA) and thus evolve quickly. The technology for mtDNA, however, came before the use of microsatellite DNA.

Finally, this strict maternal transmission and lack of recombination—the very aspects that made mtDNA unique—also made mtDNA attractive at first for tracing genealogies. Scientists like to first figure out the simplest cases before moving on to more complex ones. Darwinian detectives are no exception. It's easier to trace back the simple patterns of maternal inheritance and no recombination than it would be to examine the more complicated genealogies that nuclear DNA entail. The lack of recombination means that all the genes in the mitochondrial DNA share the same fate. From the viewpoint of an evolutionary geneticist interested in tracing genealogies, the mtDNA is one single gene.

■ The Mother of All Mitochondrial DNA

Our story starts in the mid-1980s at University of California at Berkeley in the laboratory of Allan Wilson, a prominent molecular evolutionary biologist who originally hailed from New Zealand. Wilson's first claim to fame came in the 1960s when he and Vincent Sarich had shown that humans and chimpanzees diverged not 20 million years ago as the paleontology community had been saying at the time, but less than ten million years ago. Sarich and Wilson reached their conclusions by examining immunological reactions across species. If a rabbit were exposed to a human protein, its immune system would recognize the protein as foreign, and would make antibodies against that protein. Obviously, those antibodies would respond to another human protein. But would they react to the same protein from a chimpanzee? Yes, but the response would not be as strong. By comparing how the rabbit antibodies responded to proteins from a variety of species of primates and by employing the logic of the molecular clock, Sarich and Wilson concluded that humans and chimps probably diverged five to ten million years ago. These methods, crude by even the standards of the 1980s, were the cutting edge back in the 1960s. The conclusion of Sarich and Wilson of a human-chimp split around six million years ago has been confirmed numerous times, with a variety of types of molecular and morphological data.

During the 1980s, Rebecca Cann was a graduate student working under the guidance of Wilson. By this time, mitochondrial DNA had been used to confirm the comparatively young age for the human-chimp split that Sarich and Wilson proposed. Cann, Wilson, and Mark Stoneking (another Ph.D. student in Wilson's lab) were interested in applying mtDNA to address the questions of more recent human origins. They wondered whether they could trace back the mitochondrial DNA of all humans to the most-recent common ancestor, the mother of the mtDNA. As noted previously, this individual must have been a mother and not a father, because only females transmit mtDNA.

Cann, Stoneking, and Wilson did not use the term "mitochondrial Eve" in their original paper.[6] Instead, the first appearance of this catchy but also potentially misleading phrase was by a science journalist writing about the paper by Cann et al. in the journal *Science*. The term is misleading not only because of possible Biblical connotations, but also because it seems to suggest that this person was the first human female, or even among the first humans. The mitochondrial Eve could have come well after the appearance of modern humans. It's also possible that this Eve lived before modern humans, and thus did not look like us. These claims may seem counter-intuitive, but they will be explained in the following paragraphs. Throughout the remainder of this chapter, references to Eve are to the mitochondrial Eve found with genetic methods, not the Biblical Eve.

So what do we mean when we speak of this mitochondrial Eve? It is a mathematical certainty of evolutionary genetics that we can trace the copies of any particular gene present in a species back to a common ancestor. Lineages die out every generation, and as we saw in the case of my personal family tree, mitochondrial lineages can die out quickly. Recall that my mother's mother's mother, despite having two daughters and a total of nine great-grandchildren, lacks a direct female line to continue to transmit her mtDNA. At some point in the future, all lineages of mtDNA (or any gene) existing at one time save one, eventually go extinct, leaving but one most-recent common ancestor of the gene.

Eve is just a name we give to the most-recent common ancestor of all of the copies of mitochondrial genes in a species. Such a designation can only be made retrospectively; that is, at the time Eve lived, it would be impossible to determine that she would eventually become Eve. For a considerable period of time, this person was not a common ancestor of all of humanity's mtDNA; at that time, some other woman was Eve. Moreover, at some point— probably thousands of generations from today—there will be a different Eve. Let me repeat that: assuming that our species doesn't go extinct, a different person will be considered "Eve" at some future date. Why is that? Suppose a hypothetical woman, call her Ms. X, is the most recent common ancestor for all mtDNA types except for one small lineage. Because Ms. X is not the common ancestor of that distinct lineage, she is not the common ancestor of *all* mtDNA types, and thus cannot be Eve. But if the last member of that distinct lineage dies outs or has only males, then Ms. X would become Eve, because now all mtDNAs trace back to her.

Eve was not the only female of her time who bore children. In fact, no evidence of a severe population bottleneck around Eve's time exists. Most likely, nothing is unusual about Eve, except that she had granddaughters from at least two daughters. Why must Eve need to be the mother of at least two daughters and not just one?

To answer this, consider the hypothetical case of a woman whom we shall call Lilith. I picked this name because in some mythologies, Lilith was the first wife of the Biblical Adam. The singer-songwriter Sarah McLachlan also named the Lilith Fair after this mythical Lilith. Getting back to our example: Suppose Lilith, who had only one daughter, was a common ancestor of all mtDNA. That is, all of humanity could trace their mtDNA back to Lilith. Because Lilith had only one daughter, all of humanity must trace their DNA back to Lilith's daughter, who would thus, by necessity, also be a common ancestor of all mtDNA. Thus, Lilith could not be the *most recent* common ancestor of all mtDNA. Now suppose Lilith's daughter had at least two daughters who produced granddaughters. Each of Lilith's daughter's daughters would start a separate branch, and the most recent common ancestor of all mtDNA would thus be Lilith's daughter. We could then designate her as Eve.

How did Wilson's group track back to the most recent common ancestor of our mitochondrial DNA, this Eve? They started with the idea that an Eve, a most recent common ancestor, existed, and then they worked backwards, both logically and chronologically. In this way, the method of Cann, Stoneking, and Wilson was not all that different from what traditional genealogical researchers do. Although the story of *Roots* was told chronologically from Africa circa 1767 to 1970s America, Haley's detective work was largely backwards. He started with the family he knew, and traced his way back generation by generation.

Rebecca Cann had gathered genetic data from a region of the mtDNA that had a high mutation rate and, hence, would evolve quickly. She managed to acquire DNA from 147 individuals across the world of different ethnic origins; many of these samples were from placental tissue obtained from local hospitals. Cann found many differences among the mtDNA samples and that most individuals differed from one another; in fact, there were 133 different haplotypes among the 147 samples. Starting with the supposition that a most recent common ancestor for the mtDNA must exist, Cann and her colleagues reasoned that the differences observed in the mtDNA present in humanity today must have arisen via mutations since the most recent common ancestor, Eve.

After reconstructing Eve (we'll discuss in more detail below how they did this), Cann and her colleagues then calculated the average genetic difference between Eve and the contemporary individuals in their study. This provides a measure of how much evolution had taken place at the mitochondrial DNA since Eve lived. They were able to assign approximations of dates of splits because they knew that New Guinea had been settled about 40,000 years ago, and from this information they could calibrate the molecular clock. In this way, they could establish an estimate for when Eve lived.

■ Out of Africa

Two major branches are apparent in the "family tree" (biologists call it a phylogeny) that Cann and her colleagues obtained with their mtDNA data. One branch comprises individuals who are all of African descent; the other is comprises individuals from all over the world, including Africa. The logical inference to be drawn from this pattern is that there was a split in the tree, and that split occurred in Africa. This means that Eve, the common ancestor of all of our mtDNA, lived in Africa.

Cann and colleagues also determined that Eve lived somewhere 140,000 and 240,000 years ago. This figure was based upon how many mutations had accumulated, scaled against the molecular clock. They set upon a middle-range estimate of 200,000 years old, which would be about December 19 by our human calendar, give or take a day or two.

Nothing about this date is particularly special. This estimate of Eve's appearance is consistent with the time that the modern facial features of our species appear. If the dates of Eve and the appearance of anatomically modern humans agree, however, it is likely to be just a coincidence. Eve does appear substantially before Diamond's "Great Leap Forward," the period about 50,000 years ago when cultural and technological advances began to accelerate.

The date of Eve's appearance provides us with information about the effective population size of humans as experienced by the mitochondrial DNA. If the effective population size is large—that is, many breeding females are in the population—it will take a long time for all but one of the lineages to die out. If the population has only a few breeding females, even for a short period, lineages will die out at a faster rate. Calculations based on population genetic theory suggest that the long-range effective population size of our species has been around 10,000 females. This estimate is based only on the mtDNA and thus does not tell us anything about the number of breeding males.

As we saw in chapters 3 to 5, natural selection can either increase or decrease variation, and hence effective population size, depending upon which type of selection is operating. The mitochondrial DNA is certainly under negative selection, a situation that would tend to decrease variation and hence effective population size. Positive selection has probably played some role in the evolution of mtDNA, and that, too, would tend to decrease variation and effective population size. It is *unlikely* that balancing selection has been important in mtDNA evolution. Because individuals receive only copy of mitochondria, there are no heterozygotes for mitochondrial genotypes. Thus, heterozygote advantage (a major type of balancing selection) cannot exist for the mitochondria.

■ Critique and Complications of "Eve" Studies

Criticism of the study by Cann, Stoneking, and Wilson began almost as soon as it appeared in print.[7] Among the initial concerns was that Cann and her colleagues used African Americans instead of African populations in their survey. This criticism can be rectified easily—just by including African populations, which is what Wilson's lab did in follow-up studies.

Other, more serious criticisms focused on the methodologies used by the Berkeley team to reconstruct the phylogeny, the family tree. Increasingly since the 1960s, the process of phylogeny reconstruction has been one of the most active fields of inquiry in evolutionary biology. Problems in phylogeny reconstruction and associated fields have attracted some of the best minds not only in biology but also in philosophy, computer science, and mathematics. Only the barest summary of this wonderfully intricate field will be given in this book. I apologize to its practitioners for glossing over many details.

Whether constructed using molecular characters, as Wilson's group did, or constructed with morphological characters (such as teeth or pelvis width), a phylogeny (or tree) is a hypothesis about how evolution of a trait or traits has taken place. And as we saw in chapter 3, more parsimonious hypotheses—ones that explain the data in the simplest manner—are preferable to less parsimonious ones, all things being equal. The most parsimonious tree is generally the one that requires the fewest number of changes to take place during the evolutionary history as hypothesized by that tree. All other things being equal, a phylogenetic tree that requires 27 changes to explain the evolution of a lineage is preferable to one that requires 30 changes, because the former is more parsimonious.

Reconstructing the most parsimonious phylogeny is easy when only a few samples are involved. But when lots of samples are assayed, the task can be very difficult. The number of possible phylogenies explodes with the number of samples: three trees are possible with three samples, but over 34 million are possible with ten samples and billions and billions are possible with only 20 samples. Clearly, for anything other than a tiny study, computers must be used. But for a study the size of Cann and her colleagues, even an ultra-fast computer could not determine which tree was the shortest (the most parsimonious); thus various methods were developed to aid in the search for the shortest tree. However, back in the late 1980s, these techniques were not as powerful as they would later become, nor was computing capacity as great. Much of the early criticism that Wilson's group received revolved around the methods that they used in searching for the shortest, most parsimonious tree.

Linda Vigilant, another graduate student in Wilson's group, was first author of a paper published in 1991 that also investigated the timing and location of the mitochondrial Eve.[8] Vigilant and her colleagues addressed many of the criticisms of the original study by Cann et al., including the need to use samples from people indigenous to Africa instead of samples from African Americans. They also used better methods for searching for the shortest tree. Due to the availability of PCR technology, Vigilant and her colleagues were also able to use hair to get DNA instead of placentas; this also enabled them to get DNA from chimpanzees. Having chimp samples enabled the Berkeley researchers to use human-chimp divergence in addition to New Guinea colonization in calibrating the molecular clock; both calibrations agreed reasonably well with each other. The chimp sequence also confirmed the rooting of the tree. Vigilant's results provide strong support for Eve living in Africa around 150,000 years ago, essentially the same conclusions as those reached in the original study by Cann et al. published four years previously.

Although the controversy would linger into the 1990s and several aspects of Wilson's group's studies would be challenged, other groups would reach similar conclusions. For instance, researchers at Uppsala University in Sweden used complete mitochondrial DNA sequences (all 16,000 odd nucleotides) from 53 individuals. This group found very convincing support that the mitochondrial Eve lived in Africa and that her descendants stayed in

Africa for some time. They concluded, "The three deepest branches lead exclusively to sub-Saharan mtDNAs, with the fourth branch containing both Africans and non-Africans. The deepest, statistically supported branch (NJ bootstrap = 100) provides compelling evidence of a human mtDNA origin in Africa."[9]

Let's decode that quotation. A bootstrap is a statistical test that researchers use to determine how well the data support the hypothesis represented in their phylogeny; a bootstrap score in the 70s is good support, one in the 90s is very strong support, and a score of 100 is the highest possible support. So, very strong support exists for these deep branches that include both Africans and non-Africans.

The Uppsala group placed the most likely date of the mitochondrial Eve at 170,000 years ago, very close to Cann's original estimate. Cann, Stoneking, and Wilson in their 1987 paper did use what would be considered crude methods and analysis by the standards of the early twenty-first century, and they did made several errors in that original paper. Nevertheless, the central points of their paper have been vindicated—the most recent common ancestor of the mtDNA of all humanity lived in Africa around 200,000 years ago![10]

Sadly, Allan Wilson did not live to see the vindication of the mtEve. At the age of 55, he died in the summer of 1991 while receiving treatment for leukemia. Rebecca Cann, however, continues to work in molecular evolution. Now a faculty member at the University of Hawaii, Cann and her students and postdoctoral fellows study the population genetics of peoples indigenous to the Pacific Islands.

■ Seven Daughters of Eve

The British geneticist Brian Sykes and his colleagues identified seven clusters of mitochondrial haplotypes within which over 95% of all modern people with European ancestry now fall.[11] The clusters, which are between 10,000 and 45,000 years old, had their origins in or near Europe. Just as we can trace back the mtDNA of all humanity to a single individual woman—the mitochondrial "Eve"—we can also trace back the mtDNA of all individuals in each of these clusters to a single woman. Such clan mothers, like Eve, must have had granddaughters from at least two daughters. Brian Sykes dubbed these clan mothers "the seven daughters of Eve" in a book of the same name.

These women are not really daughters of Eve, except in the sense that we are the daughters and sons of Eve, as far as our mitochondria are concerned. The "seven daughters" lived substantially closer to our time than they did to Eve's era, and all lived during or after Diamond's "Great Leap Forward." The largest cluster, in which 47% of all modern Europeans belong, is neither the oldest nor the youngest. The clan mother of this cluster (Sykes named her Helena) most likely lived 20,000 years ago, probably the time of a severe Ice Age. Why is this cluster so large? No one really knows. Perhaps some change occurred on this mitochondrial lineage that provided some evolutionary advantage. Alternatively, the dominance of this particular type could have nothing to do with the mitochondria. Perhaps, early on, a population that had a high frequency of the Helena type mtDNA came up with some "killer application" that enabled them to survive better. But maybe Helena's children were just lucky.

■ The Y-chromosome Adam

Another region of the genome, like the mitochondria, is passed down only in one sex: the Y chromosome. It is transmitted solely from father to son and is never found within a female (except when a pregnant woman carries a male fetus). The Y chromosome does not recombine with its partner, the X chromosome, which is present in both males and females. Women normally have two X chromosomes, whereas men normally have but one. It is the presence or absence of the Y chromosome (specifically, the sex-determining gene, *Sry*, on the Y) that determines whether a person will be male or female, respectively.

Largely because it does not recombine, the Y chromosome in mammals can and does accumulate genetic "junk." Containing over 40 million nucleotides, over two thousand times the length of the mtDNA, the Y chromosome has comparatively few genes, other than the previously mentioned *Sry*.

The Y chromosome has been useful in solving a two-centuries-old mystery. When Thomas Jefferson was President, a political journalist named James Callender wrote in a Richmond newspaper that Jefferson had kept one of his slaves as a concubine, fathering several children with her. Callender identified this woman as Sally, and it was clear that Jefferson had had a close relationship with a slave named Sally Hemmings. The controversy simmered from time to time but came to a boil recently when the tools of molecular biology were brought to bear on it. This scandal of 1802 was apparently solved in 1998, a time when another United States President (with the middle name of Jefferson) was involved in a sex scandal. Researchers found that direct male-line descendants of Eston Hemmings, Sally Hemmings' youngest son, had the same Y chromosome type as direct male-line descendants from Jefferson's lineage. Although Jefferson himself left no direct Y chromosome heir (he had five daughters but no sons who survived past infancy), he would have the same Y chromosome as his father's brother, who did leave direct male-line descendants. It is still formally possible that someone from Jefferson's family who had the same Y chromosome type as Thomas Jefferson fathered Eston Hemmings, but Jefferson is the most likely father.[12]

Can we find the Y chromosome counterpart—dare we call him Adam?—of the mitochondrial Eve? Finding this "Adam" has been harder than tracing back the mitochondrial Eve; in large part, this was because initially researchers had a difficult time uncovering variation on the Y chromosome.[13] In 1995, a paper was published in the prestigious journal *Science* in which no genetic variation was found in a 729 nucleotide-long region of the human Y chromosome among 38 men sampled around the world.[14] The 1990s television show *Seinfeld* claimed to be about nothing; was this report a paper about nothing? In actuality, the lack of variation among men for the Y chromosome does tell us something; it provides an upper limit for the age of the Y-chromosome Adam. The upper limit, as estimated in the study, was 800,000 years ago, but probably was much less than that.

During the late 1990s, improvements in technology allowed for easier identification of polymorphic markers linked to the Y chromosome. In 2000, Peter Underhill and over a dozen colleagues published a study that found 167 Y-linked polymorphisms clustered into 116 haplotypes.[15] They estimated that the most recent common ancestor of Y-chromosome haplotypes—the Y-chromosome Adam—probably lived just under 60,000 years ago, with a range of 40,000 to 140,000 years ago.

The Y-chromosome Adam almost certainly lived in Africa. This conclusion is based in part from the geographical distribution of one haplotype; found mainly in Ethopian, Khosian, and Sudanese populations, it never has been observed in people without recent African ancestry. Other studies of the Y chromosome also support the recent age of the Y-chromosome Adam, as well as its placement in Africa. In fact, one study of microsatellites on the Y concludes that the split between Africa and non-Africa was only 24,000 years ago![16]

The Y-chromosome Adam is almost certainly younger than the mitochondrial Eve. Though it is perhaps counterintuitive, we will see that such a finding is to be expected based on population genetics theory. Among other things, the age of a most recent common ancestor reflects the effective population size of the genetic entity. Entities with a high effective population size should have a more ancient most recent common ancestor than entities with a low effective population size. The effective population size of males (as traced through the Y chromosome) is about 3,500, much smaller than the effective population size of females (as traced through the mitochondria). Why should this be?

The reason has to do with a fundamental difference between male and female reproductive behavior in humans and most other mammals. It is true, that because every one of us has exactly one genetic father and exactly one genetic mother, the average number of offspring per male must be exactly the same as the average number of offspring per female. But the average does not tell the whole story of the distribution. Although the averages are the same, the variance of number of offspring of males is generally larger, and sometimes much larger, than that of females! Males are more likely to have either a lot of offspring or no offspring at all, than are females. This asymmetry affects effective population size. Because fewer different males than different females contribute to the gene pool, the effective population size of males is less than that of females. Another way to say that is that the effect of genetic drift on genes transmitted only by males is greater than the effect of drift on genes that are transmitted only by females. Lineages transmitted only through males should die out sooner than lineages transmitted solely through females. Thus, it is not a surprise that the Y-chromosome Adam lived more recently than the mitochondrial Eve.

In the Underhill study and in other studies of both Y chromosomes and mtDNA, researchers consistently find more variants at rare frequency than would be expected under the neutral theory. This is the opposite of the pattern found in the regulatory region of the chemokine receptor *CCR5* (chapter 5) wherein the excess of variants of intermediate frequency was taken as evidence of long-lasting balancing selection. In contrast, the excess of rare variants is consistent with a model of a rapid expansion of population size. There are other possible reasons for an excess of rare variants, so we should not say that these results prove that the human population size exploded in the past, just that these results are consistent with a population explosion. Moreover, we have many other independent lines of evidence for such a population explosion.

As with the "seven daughters of Eve," geneticists have traced back inter-mediate ancestors for Y-chromosome types between Adam and us. The vast majority, and perhaps all, men with European and Asian genetic backgrounds can trace their Y-chromosome lineage back to a particular male (named M168, after the marker that defines these chromosomes). M168 thus can be considered the Eurasian Adam. Although the Y-chromosome Adam and the mitochondrial Eve did not meet, it is quite possible that the Eurasian "Adam" M168 could have met his equivalent, the Eurasian Eve (known as L3). The estimates of their dates overlap (around fifty thousand years ago) and they both probably lived in northeast Africa. Africa? Yes, Africa. Although nearly all Eurasian mtDNA and Y chromosomes currently existing can be traced back to L3 and M168 respectively, M168 and L3 also had African descendants.

Some differences exist between the Y chromosome genealogies and those obtained from tracing other genetic entities. For instance, the Y chromosome from the Welsh population is distinct from those from the rest of British Isles, but the mtDNA is the same. The Welsh Y chromosome appears to be quite similar to the Y chromosome from the Basques. The origin of the Basques, who reside in the area between Spain and France, is still disputed. Some argue that they arrived as part of the group of Indo-Europeans who invaded Europe four or five thousand years ago; others put forth that Basque history predates the Indo-European invasion by many millennia. Either way, the Basques have been isolated for a considerable time and are quite distinct from the Welsh, so the similarity of their Y chromosomes is surprising.

The larger picture is that a gene genealogy is not necessarily equivalent to individual-based genealogy; in fact, genealogy can and does vary among different genes. Variation in genealogy means that the environments that genes experience will not be the same.

■ Out of Africa Again and Again

One of the most heated recent controversies in evolutionary anthropology has been the "out of Africa" versus "multiregional" debate. The out-of-Africa proponents, bolstered by the results from the mtDNA and the Y chromosome, claim that modern human beings emerged from Africa roughly a hundred thousand years ago, replaced the populations living elsewhere, and populated the globe. In contrast, the multiregionalist hypothesis is the proposition that modern humans arose in several geographical regions (not just Africa), and that these populations have been exchanging genes for several hundreds of thousands of years. This debate actually had been raging in the journals and the conferences of the paleontology community—sometimes quite heatedly—for years before Rebecca Cann and her colleagues published the first Eve paper.

The out-of-Africa and multiregional models for the origin of modern humans are but two extreme viewpoints of a continuum; intermediate positions are possible. Although the mtDNA and Y chromosome data have essentially rejected strong multiregional positions, these results may not be incompatible with weaker versions of the multiregional hypothesis.

In recent years, gene genealogies from X-linked and autosomal genes have informed the debate. A Japanese group led by Naoyuki Takahata re-analyzed data already published from ten X-linked and five autosomal genes, as well as the mtDNA and Y chromosome data sets.[17] In nine of the ten informative cases, the most recent common ancestor was located in Africa. Takahata's group also performed computer simulations to determine whether weaker versions of the multiregional hypothesis are consistent with the data. To explain the empirical results under a multiregional framework, one would need to posit that the number of breeding individuals in the African population was much larger than in the other populations. Other populations could have been contributors to our current gene pool, but the African population must be the major source.

Alan Templeton, one of the critics of the initial Cann and Wilson studies, has re-analyzed data that partially overlaps with the data set from Takahata's group and includes genes that are X-linked, autosomal, Y-chromosome, and mtDNA.[18] Templeton argues for a model of human evolution consisting of at least two expansions out of Africa, one that occurred between 400,000 and 800,000 years ago (as reflected by nuclear genes) and a more recent one 80,000 to 150,000 years ago (as reflected by mtDNA and Y). He goes on to suggest that the most recent out-of-Africa expansion was not a replacement event, in which the invading population wiped out an existing one. Instead, it represented interbreeding and genetic assimilation.

This chapter, and much of this book, has focused on what we can learn about the genealogies of genes. As we saw earlier, the genealogies determined from tracing one particular gene could be very different from those obtained from a different gene. The mitochondria and the Y chromosome, representing female lineages and male lineages, respectively, are but two of myriad genes for which we could trace genealogies. But what about genealogies of people?[19] Joseph Chang, a mathematician at Yale, posed the question as follows: "How far back in time do we need to trace the full genealogy of mankind in order to find any individual who is a common ancestor to all present-day individuals?"[20] The answer is quite different from what we saw in the case of gene genealogies.

Thomas Jefferson is the ancestor of many people alive today, including some of the descendants of Eston Hemmings. Kunte Kinte, of Jefferson's generation but very different circumstances, also has many descendants. Only a very small fraction of people currently around, however, could claim either Jefferson or Kinte as an ancestor. As we go back further in time, the potential increases for one person to be an ancestor of a large fraction of contemporary humanity. Indeed, at some point, a person could be the demographic ancestor of all of humanity. The actual proportion of his or her genome that is in the genomes of people alive today could be very low; he or she may not have had passed on any genes to some people alive now. Still more distant in the past, a certain point is reached where all people alive at a given time are either the ancestors of everyone living on Earth now or the ancestors of none. There ceases to be an in between. The work by mathematical demographers shows that that point is surprisingly early—well after the lives of the genetic Eve and Adam, and perhaps only a few thousand years ago. The exact time of that depends upon many factors, including how much gene flow has occurred in populations. Another fascinating conclusion of this work is that any given time prior to an "ancestor of all or ancestor of none" point, about 80% of those who had children are the ancestors of everyone alive now.

7 ■

Who Were the Neanderthals?

> *...recent research is finally establishing that the*
> *Neanderthals were fully human, but were also quite*
> *different from Homo sapiens.*
>
> —Chris Stringer (Stringer, 2002, p. 58)

■ The Neanderthal Puzzle

If you were to look head-on at a skull of a Neanderthal, what would you see first in that face?[1] Those big brow ridges? That humongous, projecting nose? The absence of a chin? Would you see a human, or something else?

Turning the Neanderthal skull around, you'd see a prominent bulge at its base. This feature is called the occipital bun because it arises from the projection of the occipital bone, a saucer-shaped bone located at the lower part of the cranium. Occipital buns are relatively rare in contemporary human populations, although they are found more frequently in some populations (including Lapps, Finns, Bushmen, and Australian Aborigines). Does the occipital bun, or the lack thereof, have a function? Could the bun just be a consequence of possessing a large brain but a narrow head? Nobody knows.

Yes, Neanderthals had large brains. Unlike the relatively small-brained australopithecines and even the early members of the genus *Homo*, Neanderthals didn't lack cranial capacity. In fact, on average, their brains were slightly larger than our own.

Winston Churchill once said that Russia was "a riddle wrapped inside a mystery in an enigma."[2] The same can be said for the Neanderthals. So many questions surround the nature of Neanderthals:

> Were they human? Did we evolve from Neanderthals? Are they a missing link, or an offshoot? Are they our father or our uncle?

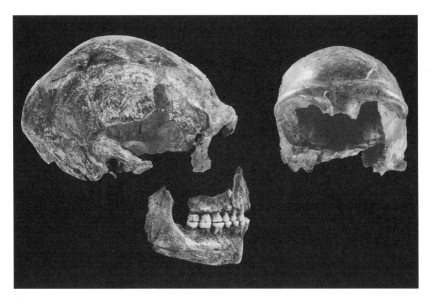

Figure 7.1
Two views of a Neanderthal skull. This fossil was found in the town of Spy in Belgium. Courtesy of Frank Williams.

Should we consider Neanderthals to be a separate species from us, *Homo neanderthalis*, leaving us as the only beings to carry the title *Homo sapiens*? Or should they be given subspecies rank, *Homo sapiens neandethalis*, leaving us as *Homo sapiens sapiens*?

What led to their decline? Did modern humans wipe out the Neanderthals as they advanced out of Africa into Europe?

Is enigma even the right word to describe the Neanderthal? The term "enigma" connotes something that cannot be explained. Puzzle may be a better word to describe the challenge Neanderthals present to those interested in our evolutionary history; puzzles, though baffling, have correct solutions.

■ The Cooking Hypothesis

In 1856, three years prior to Darwin's publication of *The Origin of Species*, workers digging in a cave near Dusseldorf and the Neander River in southern Germany discovered a skullcap and other human-like bones. The individuals to whom these bones belonged soon became known as Neanderthals; "thal" means valley in German, so Neanderthal is literally the valley of the Neander River. Though the word "thal" in the German language was shortened to "tal" at the end of the nineteenth century, Neanderthal remains the preferred spelling, though some use Neandertal. My spell checker wants to use

Neanderthal, and thus that's how it will be spelled here. In actuality, Neanderthal bones had been found prior to those located in the Neanderthal site; the first Neanderthal skull was discovered eight years previously in Gibraltar, though its significance was not recognized at the time.

The scientific community of the 1860s was not sure what to make of Neanderthals. Some saw Neanderthals as being modern humans whose bones had become pathologically deformed. Others would emphasize ape-like characteristics of the specimens, even to the extent of denying the obvious fact that Neanderthals were fully bipedal. During the century after the Neanderthal finding, Neanderthals would become neglected as anthropologists became enraptured by other fossils of "early man."

About a century after their discovery, Neanderthals would regain prominence in the eyes of anthropologists. Starting in the early 1960s, the renowned anthropologist Loring Brace[3] at the University of Michigan would promote Neanderthals as our ancestor, as the last "missing link" between ape-like organisms and us. Brace, born in 1930, was an avid follower of the modern synthesis of evolutionary genetics and had been interested in applying those ideas (the ideas of Ernst Mayr and Theodosius Dobzhansky) into evolutionary anthropology. Brace's ideas had been particularly influenced by the competitive exclusion principle, which says that two species that share the same niche cannot persist for long periods of time; they will either divide up the niche or one species will drive the other to extinction. For this reason, Brice thought that for most of the history since humans diverged from other apes, there was one species of proto-humans.

Brace proposed "the cooking hypothesis" to explain how we could have evolved directly from Neanderthals. The gist of this hypothesis is that Neanderthals, in the process of becoming better able at taming fire, started cooking more. The increased use of cooking led to less need for big teeth and powerful jaws. Because such structures are energetically costly, selective pressure existed to minimize them in the new environment—an environment brought on by cultural advances. These changes led to anatomically modern humans.

Starting around 1980, a new generation of evolutionary biologists and anthropologists challenged Brace's views about Neanderthals being our direct ancestor. Prompted by observations of the bushiness of the early stages of our lineage, these "Young Turks" wondered whether bushiness remained in the more recent stages of our evolutionary lineage. In other words, given that several species of *Australopithecus* and early *Homo* simultaneously coexisted, why couldn't Neanderthal and humans coexisted simultaneously? During the 1980s, researchers found compelling evidence for anatomically modern humans and Neanderthals being in the same general region for thousands of years.

The Neanderthals were around for a considerable period. Depending upon the authority, Neanderthals first appeared in Africa somewhere between 200,000 and 400,000 years ago. They moved into the Middle East, and by

60,000 years ago were in Europe. At the peak, the Neanderthals had an impressive range. They were all over Europe, except for the northernmost parts (such as the Scandinavian regions); they occupied large regions of southwestern Asia, from Turkey to the lands around the Caspian Sea to what is currently Iraq and Iran. Although the dates of their appearance remain controversial, less doubt remains about the timing of their disappearance. Starting around 40,000 years ago, they became less frequent, co-incident with the arrival of anatomically modern humans in Europe. By 28,000 years ago, they were gone.

■ Morphology

The morphological differences between humans and Neanderthals are striking; surely, we must be different species. Not so fast! Appearances can be deceiving. Although modern humans from different geographical regions look different, those differences are the result of a relatively small number of genes upon which different local selective pressures have operated. At the genetic level, much more variation exists within local populations than among populations.

So how do the differences between humans and Neanderthals compare with differences within humans? Sophisticated morphological studies of brain and face features find that humans and Neanderthals do differ much more than do geographic populations of modern humans.[4] These studies also took care to deliberately avoid including features that previously distinguished modern humans from Neanderthals. Moreover, the morphological difference between humans and Neanderthals was greater than those observed within species of other primates.

The morphological differences between humans and Neanderthals may be due to patterns of growth. The emerging consensus is that Neanderthals grew very quickly, much faster than modern humans do. For instance, Fernando Ramirez Rozzi and José Bermudez De Castro studied patterns of teeth development from Neanderthals as well as several other species of the genus *Homo*. Teeth grow in daily and yearly patterns; just as one can determine the age of a tree by observing its rings, researchers can examine tooth patterns under the microscope to determine the age of the teeth. Ramirez Rozzi and Bermudez De Castro found that Neanderthals formed tooth crown at a rate 15% faster than that of modern humans. The pattern seen in Neanderthals is not found in ancestral species of *Homo* or in modern humans. Ramierz Rozzi and Bermudez De Castro state that their findings not only support the notion of Neanderthals as a separate species but also provide clues about the ecological context of Neanderthals. They view Neanderthals "as a species of *Homo* adapted to particular environmental conditions, when a high-calorie diet and a high metabolic rate were able to fuel fast somatic growth, as well as to grow and sustain a large brain."[5]

Although biologists often use morphological characteristics to make inferences about whether two populations are separate species, the key criterion for species status is reproductive compatibility, the extent to which males and females from the different species will recognize each other as mates and produce successful, fertile hybrids. If the populations are more or less fully reproductively compatible, then they are considered part of the same species. If they lack reproductively compatibility and are thus reproductively isolated, then biologists generally recognize them as two separate species. There are gray areas, and not all biologists agree with basing species decisions on reproductive compatibility, but most do.

In 1999, there was a highly provocative claim of a hybrid between a Neanderthal and a modern human.[6] This claim was based on a partial skeleton of a four-year-old boy, which appeared to have a mosaic of Neanderthal and modern human features. Specifically, the body proportions and a few other characters appear Neanderthal whereas other features, including the mandible area and the teeth, resemble those of early modern humans.

Another interesting feature of this specimen is its age—24,500 years, at least a couple millennia after the disappearance of the Neanderthals. If this is a hybrid, either the Neanderthals lasted far longer than commonly believed and/or the hybrid is not a first generation (F1) hybrid but an admixture that had been maintained for a considerable period of time. The claim for this individual being a hybrid has not been confirmed. Moreover, as Tattersall and Schwartz in an accompanying commentary to the original paper point out,[7] morphological traits should not appear so dichotomous after tens or hundreds of generations of hybridization. They suggest that it is still quite possible that the fossil is just a chunky modern human child, who had descended from the invaders that kicked the Neanderthals off the Iberian Peninsula.

On the basis of morphology, Neanderthals and humans differ markedly; their differences are sufficiently large that they should be considered at least different subspecies, if not distinct species. But what can we say about the relationship between humans and Neanderthals based on DNA studies? Where do we think the Neanderthal ancestor split off the human family tree?

■ Eve and Adam and the Neanderthals

At first glance, the data from the studies of mitochondrial Eve and the Y-chromosome Adam would seem to rule out any possible affinity between modern humans and the Neanderthals. Eve and Adam's location, Africa, and timing, well after the Neanderthals' appearance in Europe, would imply that Neanderthals and modern humans didn't interbreed. But is this really the case? No, not necessarily. They still could have interbred. The Y-chromosome and mtDNA studies alone do not prove that humans and Neanderthals aren't part of one interbreeding species. It is still possible that mitochondrial variants descending from Eve and Y-chromosome variants descending from Adam got into the Neanderthal gene pool. Neanderthal genes may have entered into the gene pool of modern humans and still could be with us today. By our best estimates from the genetic data, the descendants of Adam and Eve left Africa around 50,000 years ago, well before the presence of the appearance of anatomically modern human beings in Europe. We could have been exchanging genes with Neanderthals for more than †10,000 years. Also recall that most recent common ancestors of genes from the autosomes and the X chromosome are much older than the Y-chromosome Adam or the mitochondrial Eve.

What if we could get DNA from Neanderthals? Would data from Neanderthal DNA inform us as to whether we should include Neanderthals together with us as one species, or place humans and Neanderthals in separate species? These questions are now past the realm of science fiction or philosophical musing; researchers can actually extract and sequence small fragments of DNA from Neanderthals.

■ A Primer on Primers and PCR

Before discussing the studies of Neanderthal DNA, we need to examine PCR, the polymerase chain reaction, because understanding how ancient DNA is recovered, and thus the potential problems associated with the process, requires knowledge about how PCR works. PCR is appropriately named; it is a chain reaction. In nuclear physics, chain reactions occur when more particles are created in every cycle of a reaction. This results in an exponential increase in the number of particles. PCR is a chain reaction in the same way; in each cycle of PCR, there are more DNA molecules than existed before. PCR is also selective in that it amplifies only specific DNA sequences. We will see the basis for the selectivity of PCR's amplification when we discuss its stages.

Three distinct stages compose each cycle of the chain reaction of PCR. In the first stage, the DNA is exposed to temperatures near the boiling point of water. This causes the two strands of double-stranded DNA to separate because the hydrogen bonds that normally hold the two strands together are relatively weak and will break apart at high temperatures. Ordinarily, DNA is kept at this high temperature for only a minute or two during each cycle—just long enough to ensure that all of the DNA molecules are single stranded.

We come to a critical step in the next phase of the cycle when temperature is reduced. At the lower temperature, the two strands of DNA could come together (reanneal is the technical term) because hydrogen bonds tend to form between cytosine and guanine and between adenine and thymine. If the two strands of DNA are allowed to come together, no chain reaction ensues. We would be left with the same number of DNA molecules as before. So how can we amplify the DNA we choose and only amplify that DNA?

The answer is primers; these are very short, single-stranded pieces of known DNA sequence that are added to the PCR mix before the reaction

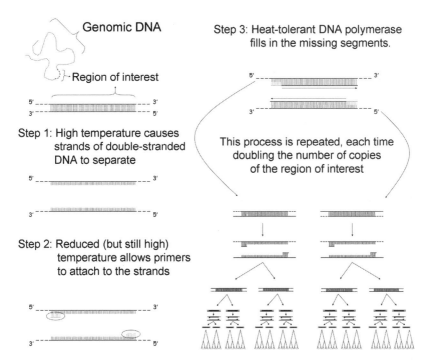

Figure 7.2
The polymerase chain reaction (PCR). The repetition of the three steps of PCR—denaturation, annealing to primers, and elongation by DNA polymerase—allows doubling of specified DNA sequences every cycle.

starts. Annealing is dependent upon the length of the DNA; short pieces anneal faster than longer ones. When a primer comes into contact with a sequence that is close to its complement (A pairs with T, C pairs with G), the nucleotides in these primers form hydrogen bonds with their complements; for instance, a primer that is AGTCTTCGGTATCGA will anneal with a sequence that contains TCAGAAGCCATAGCT. This primer is annealed to the DNA much faster than the DNA molecules can reanneal. One primer annealed to a DNA fragment wouldn't allow for amplification of the fragment. What researchers do is to use two primers at opposite ends of the DNA fragment that they want to amplify. How that amplification works is the subject of the third phase in the PCR cycle.

After sufficient time has passed for the two primers to anneal at opposite ends of the desired DNA fragment, the temperature is raised to a point high enough to stop unwanted annealing but low enough to keep the primers attached to the desired DNA fragment. Then DNA polymerase,[8] an enzyme that cells use to replicate DNA, comes in and fills in the missing DNA nucleotides between the two primers. This results in an intact, doubled-stranded DNA fragment with the desired sequence. In actuality, there are two such double-stranded DNA molecules; recall that in the first phase of the PCR cycle, two singled-stranded molecules arose from the breaking apart of the DNA. In each cycle of PCR, the number of desired DNA molecules doubles as a result of these three phases: breaking into single strands, primer annealing, and DNA polymerase filling in the nucleotides between the primers.

PCR can amplify selectively because the primers are specific in where they anneal; they will much more easily anneal to fragments of DNA with their exact complement. But the match need not be exact. For instance, the primer AGTCTTCGGTATCGA will anneal most readily to its exact complement TCAGAAGCCATAGCT, but it also can anneal to TCAGAAACCATAGCT, with less ease. The temperature at which the annealing is done will determine the extent to which annealing can occur with imperfect matches. When it is critical to get the exact match, the annealing phase should be done at a relatively high temperature. If, on the other hand, the researchers are less concerned about getting an exact match and actually want to pick up matches that are relatively close, they would then do the annealing step of PCR at a relatively low temperature.

■ Neanderthal DNA

By the 1990s, it had become possible to get DNA from fossils that were a few thousand to tens of thousands of years old. Such DNA is of small fragments, as the larger fragments have degraded. Owing to its greater abundance, mtDNA is much more likely to be recovered than nuclear DNA. It is currently impractical to get DNA fragments from fossils older than about 100,000 years, but the most recent Neanderthals lived well before that time.

By 1997, the first report of Neanderthal DNA was published.[9] Svante Pääbo, a Swede who had been a graduate student in Allan Wilson's lab, has been interested in both the evolutionary history of the human lineage and in the recovery of ancient DNA. Now at the Max Planck Institute the Max-Planck Institute for Evolutionary Anthropology in Germany, Pääbo has led the efforts to acquire DNA sequence from Neanderthals. Matthias Krings was the first author of the 1997 report, which also included Mark Stoneking, who had been a coauthor on the first mtDNA Eve paper. Krings, Pääbo, Stoneking, and their colleagues acquired mtDNA from the original 1856 Neanderthal site, where the bones are about 40,000 years old.

Pääbo's team used various primers to amplify a region of the mtDNA that evolved quickly. Because the DNA was so degraded, these researchers could only amplify small (approximately 100 nucleotide) pieces of DNA at a time. But by having these small pieces of DNA overlap one another, they were able to put together 378 nucleotides of continuous sequence from the Neanderthal. Concerned about contamination from modern human DNA on a specimen that had been handled for 140 years, this team took special precautions against that possibility.

Their Neanderthal sequence was compared with 994 contemporary human sequences. The average difference between humans and Neanderthals is 27 nucleotides—more than three times the average number of differences between any two of the modern human sequences, which is 8. Moreover, the difference between the Neanderthal and the human DNA didn't vary according to the geographic locale of the human DNA; peoples native to Europe, Asia, Africa, the Americas, Australia, and Oceanic regions all differed from the Neanderthals by roughly the same amount. Far less divergence occurred between human and Neanderthals than between humans and chimpanzees. Krings and his colleagues also estimated that 600,000 years separate humans and Neanderthals; their estimate for the time of mitochondrial Eve of humans was 135,000 years ago (within the range of other estimates) in Africa.

The findings by Krings, Pääbo, and their colleagues were quite controversial. It is possible, but unlikely, that what they picked up was not Neanderthal DNA but contamination. Moreover, the individual fragments of DNA they obtained were quite small; they got 378 nucleotides by putting the pieces together. Another problem was the track record of ancient DNA studies themselves. Although many of the ancient DNA studies have been replicated, others have not. As Matthias Hoss pointed out in 2000, "several of the most spectacular claims—such as the retrieval of DNA sequences from 15-million-year-old plant compression fossils from 80-million-year-old bones of putative dinosaur origin and from insects of up to 130 million years in age trapped in amber—could not be reproduced in any other than the original laboratories, and so are of limited value."[10] Despite the technical expertise associated with Pääbo's group, many scientists felt nagging doubts about the Neanderthal DNA findings.

Those doubts were alleviated three years after the report of Pääbo's team when a Russian group, led by Igor Ovchinnikov, published their recovery of mtDNA from a Neanderthal.[11] This Neanderthal was found in the Mezmaiskaya Cave in the Caucasus Region of Russia. It also lived at a different time—29,000 years ago, just prior to the time when Neanderthals disappeared. The confirmation that DNA could be recovered from a different Neanderthal from a different locale by another laboratory is in itself important. Repeatability and confirmation lie at the foundation of science. A finding that cannot be repeated, or is idiosyncratic to one laboratory, has far less value than one that can. Confirmation is especially important when dealing with particularly significant or sensational claims.

Ovchinnikov and his colleagues didn't just confirm that DNA could be extracted and sequenced from Neanderthals, they also showed that the Neanderthal found in Russia also had a DNA sequence that differed from those of modern humans. Moreover, its sequence resembled that of the Neanderthal found 1,500 miles away in Germany.

Another important series of results came from DNA from anatomically modern humans that lived during roughly the same period as the Neanderthals whose DNA has been examined. David Caramelli at the University of Florence and his team sequenced 360 nucleotides of mtDNA each from two individuals exhumed from the Paglicci cave in Southern Italy. One individual was dated as being 23,000 years old and the other 25,000 years old; thus, both lived soon after Neanderthals disappeared. Special care and a series of controls were used to preclude contamination. The mitochondrial DNA sequences from the individuals found from the Pagilicci cave fall "well within the range of variation of today's humans,"[12] but are very different from the Neanderthal sequences.

With respect to these old but modern-looking fossils, there is a technical concern about contamination. Are the DNA sequences that appear similar to contemporary human sequences actually from the fossil itself or from the people handling the fossil? As a control, Pääbo's group set off to examine that possibility and examined more Neanderthal and early human bones as well as cave bear teeth. They found that under some conditions, human mtDNA could be amplified from just about anything; these were obviously the result of contamination. But under other, more stringent conditions, only Neanderthal DNA would be amplified from Neanderthals. They would also find results confirming those of Caramelli and his colleagues; under the same conditions, the sequences from anatomically modern humans from 20,000 years ago looked like those of contemporary humans, but Neanderthals sequences were different.[13]

The absence of Neanderthal mtDNA in five early modern humans means that we can say with confidence that the gene pool of early modern humans could not be composed of more than 25% Neanderthal genes. To be able to exclude Neanderthals making 10% of the mtDNA, we would need to examine at least 50 early modern humans.

Mathias Currat and Laurent Excoffier from Switzerland took a modeling approach to estimate the maximum number of successful matings between early modern humans and Neanderthals that would be consistent with less than 25% of Neanderthal mtDNA existing in early modern human's gene pool.[14] They found that this maximum number was around 120. At first glance, this may sound like a considerable amount of hybridization; but if we assume that Neanderthals and modern humans coexisted in the same place and time for over t10,000 years, it is actually very little hybridization. If such a low number of hybridizations occurred over such a long time, it would severely challenge the view that humans and Neanderthals are the same species.

Currant and Excoffier made a number of important assumptions in their model, including a range expansion of modern humans from the Middle East into Europe driven by their increasing population size. They further assumed that Neanderthals competed with modern humans, and modern humans replaced Neanderthals because the former, being better able to exploit resources, could reach higher population sizes.

One interesting result from this model is that the Neanderthal genes are more likely to survive when they come into an invading and expanding population of *Homo sapiens* than they would if the population had been steady in size. Why would this be so? Think of a single genetic variant being introduced into a population (as would be the case if Neanderthals and humans rarely mated). Such a variant would be likely to be lost in the new population, just by chance. Although accidental loss would still occur if the variant were advantageous, the loss would be less likely than if the variant were neutral. Now consider that the population is expanding; in this case, each mating pair produces more offspring, and hence a rare variant has a greater chance of making it to the next generation and subsequent generations. The adage "a rising tide lifts all boats" may or may not apply in economics, but a rising gene pool will lift genetic variants.

One factor that this modeling study did not examine was the effect of selection. Suppose that the Neanderthals possessed mitochondrial genes that were slightly less fit than the mitochondrial genes of anatomically modern humans. Even a small difference in fitness would lead to Neanderthal genes getting wiped out, even if some interbreeding took place.

This caveat aside, the evidence does appear consistent with near complete reproductive isolation between Neanderthals and modern humans. Such reproductive isolation could have arisen from behavioral differences between the two groups, from hybrids having low survival or mating success, some combination of both, or from other factors.

■ Population Genetics of Neanderthals

But what of the Neanderthals themselves? Can we say anything about their demography, origins, and evolution based on DNA evidence? As of early 2006, published mtDNA sequences are available from at least nine different Neanderthal individuals. Although this is still a very small sample, it does allow researchers to start investigating the population genetics of Neanderthals themselves.

One marker is highly polymorphic among the Neanderthal DNA sequences; at position 16258, four of the nine samples have A, and five, G.[15] Modern humans are A at this site, which is assumed to be the ancestral state. There is no parallel in modern European mtDNA sequences for this level of polymorphism. Moreover, this site is not very mutable in modern human populations. The G substitution is found in Neanderthal sequences in both the Iberian Peninsula and the Balkans. This suggests that the polymorphism existed before the Neanderthal's retreat to southern refugia at the peak of an ice age about 130,000 years ago. Other estimates also point to the polymorphism being old. The estimate of the effective population size of Neanderthals based upon the mtDNA sequence is about 7,000 females—comparable to that of modern humans.

■ Conclusions and Unanswered Questions

The morphological and genetic data gathered thus far provide fairly strong support for humans and Neanderthals being separate species.[16] Some caveats about the genetic data persist, because the sample sizes are still very small both for the Neanderthals and for the early modern humans. Finding even one individual Neanderthal that has mtDNA similar to those of contemporary humans would alter some of our conclusions. Finding mtDNA that looks like Neanderthal mtDNA in the humans from 20,000 years ago would also cause our conclusions to change. Still, multiple lines of evidence are all consistent with Neanderthals being at least mostly—if not completely—reproductively isolated from modern humans. For this reason, we should award them full species status as *Homo neanderthalis*, at least on a provisional basis.

Although of extraordinary interest, the discovery of one or even a few confirmed hybrids between humans and Neanderthals would not necessarily imply that we are the same species. The reproductive isolation criterion for species is not absolute. Even discrete entities that biologists consider "good species" can hybridize occasionally but remain separate as long as the extent of gene flow between them is sufficiently low.[17]

The discovery of *Homo floresiensis* means that three species of *Homo* lived simultaneously as late as 28,000 years ago. Being only 18,000 years old, the most recent *Homo floresiensis* are well within the range from which researchers have successfully obtained mtDNA. Were such DNA to be recovered, the differences between it and modern humans would almost certainly be greater than between modern humans and Neanderthals. The presence of *Homo floresiensis* as recently as 18,000 years ago also must give us pause regarding when *Homo neanderthalis* went extinct. Just because we haven't observed any Neanderthals after about 28,000 years ago doesn't discount the possibility that they persisted longer.

So it appears that *Homo neanderthalis* and modern humans lived in the same general location for at least 12,000 years, maybe longer. How could the competitive exclusion principle account for such a long period of coexistence between two species? One possibility is that Neanderthals and modern humans divided the niche. For instance, Neanderthals may have been better hunters of some game whereas modern humans may have been better tenders of fire and superior cooks. Both Neanderthals and early modern humans appear to have lived in small groups; such population subdivision also helps to promote coexistence.

Neanderthals weren't us, but were they fully human as Chris Stringer claimed in the opening quote? What does fully human mean? The brains of *Homo neanderthalis* were at least as big as our brains. Although we do not know how their brain was wired, it seems perversely *sapiens*-centric not to assume, without evidence to the contrary, that they were reasonably intelligent—and perhaps as intelligent as we are. Although Neanderthals did not take the Great Leap Forward, neither did we until 50,000 or so years ago. Did Neanderthals bury their dead? Most of the evidence supports that they did, but some skeptics have questioned the solidity of that evidence. Did Neanderthals speak? Did they use complex language? Anatomical evidence suggests that *Homo neanderthalis* had the capacity for at least the rudiments of speech, if not much more. Perhaps we will never fully answer these questions, but we will discuss what genetic studies can tell us about Neanderthal language in chapter 8.

■ Hold the Presses

Very recent improvements to radiocarbon dating procedures may have significant implications for the Neanderthal enigma. Radiocarbon dating of artifacts more than about 20,000 years old can be error-prone for reasons that we will see below. Radioactive carbon-14, like all radioactive material, decays at a constant rate dependant upon how much of it is present. The rate at which different radioactive elements decay, however, varies and is specific to the type of element. Phosphorus-32 decays quickly; half is gone after 14 days. Uranium-238 decays very slowly; only half of the amount of uranium-238 present at the time the Earth formed approximately 4.5 billion years ago has decayed, the other half is still around. The rate of decay of carbon-14 is intermediate; half will decay every 5,700 years.

Because half of carbon-14 decays every 5,700 years, after 11,400 years (two half-lives), only one quarter of the original radioactive carbon-14 is left. After five half-lives (28,500 years), only 3.1% of the radioactive carbon remains. Six half-lives is 34,200 years; at that point, the amount of carbon-14 has been reduced to only 1.5% of the original. Dating of materials in this time range is still possible but is more subject to error and biases. Two major sources of bias can enter into assessments of the ages of artifacts that are over about 15,000 years old. The first is that even relatively small amounts of contamination of modern carbon can make the sample appear younger than it really is. In addition to the contamination problem, another source of bias is that the proportion of carbon-14 in the atmosphere has not been completely stable but instead has varied slightly over the last 50,000 years. The variability in the proportion of the radioactive carbon-14 means that estimates of age may be biased.

In a February 2006 review in the journal *Nature*, Cambridge University archaeologist Paul Mellars discusses the new advances in radiodating and their implications for the Neanderthal enigma.[18] In recent years, advances in both reducing the extent of contamination and estimating more accurately the proportion of carbon-14 throughout the last 50,000 years have provided better calibration for dating materials in the 25,000 to 40,000 year range. Based on these new calibrations, it appears that anatomically modern humans spread through Europe more quickly than previously believed. These same revised calibrations also suggest that the last Neanderthal existed further in the past than had been thought. In summary, these recent results point to a reduced period of coexistence of modern humans and Neanderthals.

Pääbo's team continues to work on studying Neanderthal DNA. Going beyond small fragments of mitochondrial DNA from Neanderthals, the team has now sequenced large parts of nuclear Neanderthal DNA. In November 2006 (during the production of this book), the team published that they had sequenced one million nucleotides of nuclear Neanderthal DNA.[19] Stay tuned!

Are We the Third Chimpanzee?

A zoologist from Outer Space would immediately classify us as just a third species of chimpanzee, along with the pygmy chimp of Zaire and the common chimp of the rest of tropical Africa.

—Jared Diamond (Diamond, 1992, p. 2)

■ The Lesser-Known Chimpanzee

What is the basis for the claim that we are more closely related to the two species of chimpanzee than we are to all other species? Even granted the validity of this claim, would Diamond be correct? Would an alien zoologist classify us as just another chimpanzee? What can we learn about own our species from sequencing the genome of the chimpanzee? How do we differ from chimps with respect to demography and other aspects of population genetics?

These and other questions will be addressed in this chapter. But before delving into the details of molecular evolution, let's get acquainted with the bonobo—the ape formerly known as the pygmy chimpanzee.[1]

Far less is known about the bonobo than what is now called the "common" chimpanzee. Part of the reason that we have lacked knowledge about this ape is that bonobo research got off to a late start. Once called the pygmy chimpanzee, the bonobo was only discovered in the late 1920s. For decades afterward, it was thought to be just a subspecies of the chimpanzee. (For the remainder of the book, I'll refer to bonobos as *bonobos* and chimpanzees as *chimpanzees* or *chimps*, except when I make clear that I am speaking about both species of chimps.) Another reason bonobos have received less attention is their lack of a charismatic spokesperson. Although several highly respected primatologists—most notably, Franz de Waal—have closely studied bonobos, none of these bonobo researchers possess anywhere near the star power of

Jane Goodall and Dian Fossey, respectively, the champions of chimpanzees and gorillas. Studies of the bonobo's sexual habits, which we will soon discuss, have led to this species' receiving much more scientific study and even notoriety in popular circles in recent years.

The bonobo is found in the wild only in the forests of central Zaire, now known as the Democratic Republic of Congo. Even there, its range is spotty and discontinuous. Like the common chimpanzee, the bonobo is an endangered species; its habitat is threatened and it is still being hunted for food and the pet trade. In fact, hunting of bonobos has increased recently with the intensification of the civil wars in Zaire. Although no one is certain of the exact numbers, most primatologists estimate that less than 20,000 are left.

Their nickname, "the pygmy chimpanzee," is a poor descriptor because bonobos are only slightly smaller than common chimpanzees. Bonobo males generally weigh close to 100 pounds; females are somewhat smaller (around 80 pounds). Although similar in size, bonobos and chimpanzees do differ substantially in build; the former species is more slender and longer limbed. As de Waal put it, "in physique, a bonobo is as different from a chimpanzee as a Concorde is from a Boeing 747."[2] The species also differ markedly in behavior. Unlike chimpanzees, bonobos don't appear to hunt monkeys. Although they will eat just about anything, bonobos focus their food gathering mainly on fruit. Watchers observing the two species claim that bonobos are more imaginative and sensitive than chimpanzees.

Both sexes of bonobos engage in sexual practices that probably could not be shown on basic cable TV, much less on PBS. For instance, male bonobos practice a ritual called "penis fencing," which consists of two males hanging "face to face from a branch while rubbing their erect penises together."[3] Females are receptive to sex basically all of the time, not just during certain times in their estrus cycle. Sex acts in this species are typically very short— the average copulation lasts only 13 seconds!

Researcher studying bonobos believe that this frequent sexual activity is a means to minimize conflict. Groups of bonobos are characterized by cycles of breaking apart and coming together. In these fission-fusion cycles, the males are usually sedentary, whereas the females move. When a female bonobo moves into a new group, her means of entry almost always involves a sexual act. Moreover, intergroup conflict in bonobos is rare. This is in sharp contrast to the common chimpanzees, in which intense intergroup conflict and fights among males for dominance are common.

A study by Amy Parish shows how different these two species are in response to controlling access to food.[4] She presented the same situation to a group of chimpanzees and a group of bonobos, giving them access to a simulated termite mound that was filled with goodies—at least what chimps and bonobos would perceive as goodies. The reactions of these two species could not have been more different. In the case of the common chimpanzee, one dominant male was able to monopolize all of the resources, meting and doling out goodies as he chose. Female bonobos, however, controlled access to the food; sexual encounters between females appeared to facilitate the maintenance of order as they fed.

■ How Do We Know We Are the Third Chimpanzee?

Until the middle 1990s, convincing evidence for placing humans and chimpanzees as each other's closest relatives did not exist. Although most early genetic and other evidence did group humans with chimpanzees (with gorillas as an outgroup), reasonable doubt still existed within the scientific community when Jared Diamond published *The Third Chimpanzee* in 1991 as to whether Diamond's claim about the relationships was accurate.

It has been generally accepted since the start of the twentieth century that chimpanzees, gorillas, and humans are more closely related to one another than they are to any other organisms, but the relationships among the three species were problematic.[5] Of these three species, three patterns of relationships are possible. First, the prevailing view—and Diamond's claim—humans and chimpanzees are each other's closest relatives, and gorillas are the outgroup. The alternative possibilities are that humans and gorillas are each other's closest relatives, and chimpanzees are the outgroup; or that gorillas and chimpanzees are each other's closest relatives, and humans are the outgroup.

Since the 1980s, evolutionary geneticists recognized that attempts to determine the relationships of closely related species could be complicated by the fact that genealogies produced by examination of different genes can differ. We saw before in the case of the major histocompatibility complex (MHC) that some human genetic variants are more similar to some chimp genetic variants then they are to other human genetic variants (chapter 5). Such polymorphisms that transcend species boundaries are generally the result of the persistent action of balancing selection. In the absence of strong balancing selection, one would not expect to observe patterns like that found in MHC. Yet different genes vary considerably in the amount of time that is required for all variants to trace back to the most recent common ancestor. Such variation is likely to result in part from differing selective regimes of the genes but also in large part from the random (what mathematical types sometimes call stochastic) processes of mutation and random genetic drift. Similar stochastic deviations can also cause differences in the phylogenies of species obtained by using different genes.

Consider the general case in which species A splits off from the ancestor of species B and C shortly before B and C separate from each other. We can designate the splitting off of A as node 1 and the splitting of B and C as node 2. Suppose a polymorphism exists in the ancestral species in which at a specific position, some individuals possess the nucleotide G and others possess the nucleotide T. If that polymorphism persists past node 2, then it is possible that species A and species B may fix nucleotide T and species C may fix nucleotide G. Such an event would make it appear that species A and B are each other's closest relatives, with C as the outgroup, despite the fact that B and C are actually the closest relatives in the true species phylogeny. The gene tree—the phylogeny established from that particular gene—is not wrong. It accurately reflects the history of that gene. The problem is that the history of the particular gene may not reflect the history of the genome as a whole, and thus the gene tree can lead us astray from determining the true phylogenetic relationships of the species (the species tree).

The problem that these polymorphisms existing in the ancestral species (and hence, called ancestral polymorphisms) can cause usually presents itself when the distance between nodes 1 and 2 is relatively small.[6] When the difference between the two nodes is large, there is enough time for one or the other genetic variant to become fixed in the ancestor of species B and C, and thus both B and C (the two closest related species) will have the same variant. This time is usually on the order of several times the effective population size in generations. If the effective population size is 20,000, then after a 100,000 generations or so, the problem is mitigated; 100,000 generations in an ape, however, is on the order of a million years. (If the effective population sizes of humans and the nonhuman apes were greater, then this problem would be worse because more variants would have been maintained during the time between nodes 1 and 2.)

So how do population geneticists cope with this pesky problem raised by the assortment of ancestral polymorphisms? The usual approach is to collect more data—that is, more DNA sequences from as many genes as possible—and then determine whether a consensus emerges among the genes. With the increasing speed and decreasing cost of sequencing, this brute force method has become progressively easier. In well-studied groups such as the great apes, such power became available by the early and middle 1990s. As the technology progresses, sequencing multiple genes will become par for the course even in groups of species from organisms that are not as well studied.

Returning to the human-chimp-gorilla story, the preponderance of genes examined does support the close relationship between humans and chimps. During the middle 1990s, Maryellen Ruvolo examined the data from a number of genes.[7] She first lumped genes that were very closely genetically linked into a single data set, as linked genes may not be independent because they tend to share the same evolutionary history and hence genealogy. As an extreme example, all of the mitochondrial genes count as one independent gene because the lack of recombination ties their fates together. She also threw out those genes that failed to conclusively support any phylogenetic hypothesis. After this culling of the data, Ruvolo was left with 14 independent genes that each provided conclusive phylogenies. Eleven of these genes support a tree with humans and chimpanzees as the closest relatives, two support a tree that places gorillas and chimps together, and just one gene supports a tree that has humans and gorillas as closest relatives. How strong is the support for humans and chimpanzees being closest relatives? Very strong; if humans and chimpanzees were not each other's closest relatives, the likelihood of obtaining so many genes supporting such a relationship and so few supporting alternative patterns just by chance is less than two parts in a thousand. Subsequent studies strongly reinforce Ruvolo's conclusion; humans and chimpanzees are each other's closest relatives.

■ Chimp Population Genetics

As with other aspects of their biology, the population genetics of common chimpanzees is better known than that of the bonobo. Although genetic variation in common chimpanzees is higher than it is in humans, the extent to which chimps have more variation depends upon the specific genes and the specific chimps sampled. Unlike humans, in whom little genetic differentiation occurs among populations, chimps are highly genetically differentiated. So much differentiation exists among chimpanzee populations that biologists now recognize three subspecies of chimps: the western African chimpanzees *(Pan troglodytes verus)*, the central African chimpanzees *(Pan troglodytes troglodytes)*, and the eastern African chimpanzees *(Pan troglodytes schwein-furthii)*.[8]

In the late 1990s, Svante Pääbo's group undertook the most intensive study of chimp population genetics as of that date. They examined over 10,000 nucleotides of an X-linked noncoding region.[9] The average difference between any two chimps is about four times that of humans, similar to the pattern that had been previously observed with mitochondrial DNA. These data illustrate that the long-term average effect of genetic drift on this X-linked gene has been greater in humans than chimpanzees; the estimated effective population size of chimpanzees is 35,000, whereas that of humans is only about 8,000. The age of the most recent common ancestor of this gene in chimpanzees is over two million years ago. By contrast, the most recent common ancestor for these X-linked sequences in humans is almost 700,000 years old. (Note that this figure is far greater than the estimated age of the mitochondrial Eve—the most-recent common ancestor varies considerably among genes!) In yet another example of how different genes can have different natural histories, the central chimps were the most variable subspecies of chimp at this X-linked locus, whereas earlier results of the mtDNA studies had shown that the western chimpanzees were the most variable.

Pääbo's group was also able to obtain some information on the population genetics of the bonobo. This species is far less genetically variable than the common chimpanzee. Based on these same X-linked sequences, the effective population size of the bonobo was under 5,000, even lower than that estimated for humans. Of course, the census population size of humans is several hundred thousand times higher than that of the bonobo.

Mutation, genetic drift, and natural selection will cause isolated populations to diverge from each other. Gene flow, however, if present, will make the populations be less genetically differentiated. The prospect of gene flow can make it difficult to decide between two scenarios: two populations that recently diverged and are no longer exchanging genes versus populations that diverged a long time ago but still maintain a trickle of gene flow between them. Jody Hey at Rutgers University and his colleagues recently developed models that allow one to determine simultaneously how long ago two populations or two species diverged from each other and how much gene flow is occurring between them. With his graduate student Rong-Ling Wong, Hey applied these methods to chimpanzees and bonobos.[10]

Wong and Hey's analysis of a total of 50 genes supports a model of evolution in which bonobos and chimpanzees are not currently exchanging genes and have not exchanged genes since their split, which most likely occurred 850,000 years ago. The estimate of the separation date is within the same ballpark as the best estimates for the time when humans and Neanderthals diverged.

In contrast to the lack of gene flow between chimpanzees and bonobos, Wong and Hey's analysis does support a low level of gene flow between the subspecies of chimpanzees. That gene flow, however, is only in one direction. On average, there is an estimated one migrant every four generations from *Pan troglodytes verus* (western chimpanzee) into the gene pool of the central chimpanzee *(Pan troglodytes troglodytes)*. In contrast, no migration occurs from the central chimpanzee to the western chimpanzee. Wong and Hey estimate that these subspecies diverged about 400,000 years ago. They also estimate that the effective population size of the central chimpanzee is around 25,000 breeding individuals, but that the other subspecies and bonobos have effective population sizes on the order of 10,000.

■ The Chimp Genome Project

Only three years after writing a "white paper" that argued that the National Human Genome Research Institute should financially support the sequencing of the chimpanzee genome, an international consortium achieved their first major goal: publication of the "draft" genome sequence. This "draft sequence," along with several supporting scientific articles and commentaries, was published on September 1, 2005, in a special issue of *Nature* dedicated to the chimpanzee genome.[11]

Although what has been published is only a "draft" sequence subject to further refinements, the work of the consortium represents a large advance in what we know about the chimpanzee genome and the differences in the genetic makeup between humans and chimpanzees. The vast majority (98%) of the genome from the draft sequence should have an error rate on the order of one in 10,000, which is sufficiently accurate to use in sequence comparisons.

Based on the sequences from whole chimpanzee and human genomes, the single nucleotide divergence between the species is 1.23%; after adjustments for polymorphisms within each species, this figure is 1.06%. This is consistent with what had been found with analysis of single genes. But what is somewhat of a surprise is the extent of divergence due to insertions and deletions— genetic material that gained or lost by one or the other species since humans and chimps diverged from a common ancestor. Both the human and the chimpanzee genomes contain between 40 to 45 million nucleotides of DNA sequence found in just one but not both species. This total divergence (almost 90 million nucleotides) represents 3% of the genome. So how much do chimps and humans differ? On the basis of changes at just the nucleotide sites, they differ by just over 1%. If instead one also examines insertions and deletions, that divergence level climbs to 4%.

The rate of nucleotide divergence varies considerably in different regions in the genome. Some, but not all, of this variation stems from differences among chromosomes. A two-fold difference exists in the divergence rate between the X chromosome sequences (changed the least) and the Y chromosome sequences (changed the most). The divergence rates for autosomes do differ among themselves, but they all fall in between those for the X and Y chromosome. What could explain such a pattern? A logical reason, one that the consortium authors suggest, is that the pattern emerges due to more mutation taking place in the male germ-line than in that of females. This phenomenon, known as male-driven evolution, occurs because males make more sperm than females make eggs. More cell divisions are thus needed in the development of sperm than in egg development. Due to these extra cell divisions and more rounds of copying DNA, more mutations occur in the male germ-line. Recall that Y-chromosome genes are present only in males and are transmitted only from father to son. Due to their male-limited transmission, Y-linked DNA sequences would be expected to mutate faster than would the autosomes. This should lead to a higher rate of evolution than in autosomes, which are present equally among females and males. By a similar logic, the X-chromosome genes should experience lower rates of mutation and divergence because the average X chromosome is present twice as often in females (which have lower mutation) than in males. Based on the differences between the rates of evolution of the chromosomes, the authors estimate that three to six times as many mutations occur in the male germline for every one that occurs in the female germline.

The chimpanzee genome study also provides insight into the patterns of evolution of mobile genetic elements, reinforcing the notion that the genomes of mammals are not static. For instance, since the split between humans and chimps, 7,000 new copies of one of these elements *(Alu)* were inserted into the human genome. By comparison, only 2,300 new *Alu* elements were inserted into the chimp genome. Given a six-million-year divergence time, these figures imply that one new *Alu* has entered our genome on average every 850 years for our genome, and once every 2,600 years for the chimp's. We will discuss more about *Alu* and other mobile genetic elements in chapter 12.

From an evolutionary genetic perspective, perhaps the most interesting aspect of the chimp genome is the insight that it can provide us about how various forms of selection operate in apes at the level of the whole genome. Knowing the differences between the human genome and chimp genome enables evolutionary biologists to calculate the overall levels of functional constraint and hence the extent of negative selection that has acted on our species. As we have seen several times, the ratio of divergence at the sites that change amino acids (replacement) to that of sites that do not change amino acids (silent) provides a rough estimate of the functional constraint of a gene. What the genome sequences provide is a global estimate of the intensity of negative selection operating on the genome. This ratio for the human-chimp divergence is 0.23; the rate of divergence at replacement sites is just under one quarter of that for silent sites. If we assume that no positive selection is occurring and that no negative selection operates on the silent sites, this result suggests that more than three-quarters of amino-acid changing sites are under fairly strong negative selection. In actuality, this figure probably underestimates the degree of constraint operating on the sites that change amino acid, because a substantial fraction of the silent sites themselves are actually under weak negative selection. Negative selection is a pervasive force on the protein-coding regions of the human genome.

Although an overwhelming majority of the replacement sites between humans and chimpanzees are under negative selection, they appear to be less constrained than replacement sites in rodents. In other words, negative selection seems even more potent in rodents than in primates. Consider the just over 700,000 genes for which copies have been identified in all four species humans, chimpanzees, mice, and rats. The ratio of replacement to silent site divergence between chimps and humans in these genes is 0.20. (This ratio is slightly less than the 0.23 ratio found across the genome; the difference probably reflects the fact that the genes with recognizable counterparts in all of these species are more apt to be more constrained.) For the divergence between mice and rats, however, the same ratio is only 0.13; thus negative selection appears more potent in rodents than in primates. One plausible explanation for the increased constraint from negative selection operating in rodents is that rodents generally have higher effective population sizes than do apes. When effective population sizes are high, selection—both positive and negative—is more efficient relative to genetic drift. Mutations that are slightly deleterious have a greater probability of becoming fixed in species with low effective population sizes than they are in species with higher effective population sizes. Other data support rodents having a larger effective population than humans do. One caveat: comparing humans and chimps with mice and rats is not the most appropriate comparison because the absolute divergence between mice and rats is considerably larger than that between humans and chimps. If the intensity of negative selection fluctuates over time, the ratio of replacement to silent sites may vary with the divergence time between species. This problem is likely to soon be moot; it is likely that in the near future, genomes will be sequenced from other primates that are about as diverged from humans as mice are to rats, and thus better comparisons will be available.

In the next chapter, we'll return to the inferences about the action of selection that the data on chimp genome sequences allow us to draw; there we will focus on positive selection.

■ We're About 98% Chimp—What Does That Mean?

"Sameness/otherness is a philosophical paradox that is resolved by argument, not by data. Genetic data can tell us precisely what we already knew, that humans are both very similar to and different from the great apes."[12] So speaks Jonathan Marks in his provocative book *What It Means to Be 98% Chimpanzee*. Although I disagree with several statements in Marks' book— for instance, Marks thinks Neanderthals should be classified as subspecies of *Homo sapiens*—I agree with his preceding statement.

Long before the sequencing of the human and chimpanzee genomes, long before the sequencing of any genes, long even before the discovery of Mendelian genetics, biologists knew that humans and chimpanzees were closely related to one another. This relationship is based upon similarities at a huge suite of phenotypic (observable morphological, physiological, embryological, and other) characters. We didn't need DNA to tell us that we are closely related to chimpanzees.

DNA sequence information has thus provided excellent confirmation of what we already knew. That is not a trivial contribution; confirmation of findings from disparate sources and methodologies lies at the heart of science. We know a great deal about relationships based on phenotypic characters. Likewise, we know a great deal about these relationships based on genetic data. And although these data sets occasionally conflict, much more often than not they reinforce each other.

In general, genetic information provides some advantages over the data based on morphological or other phenotypic characters in determining the phylogenies of close relatives. Genetic data easily rejected the hypothesis that humans are more closely related to orangutans than chimpanzees (a position held by a minority of anthropologists during the middle of the twentieth century). After a considerable effort, genetic researchers were able to show that chimpanzees and humans are indeed more closely related to each other than either species is to gorillas.

Unlike the information gathered from morphological and other phenotypic traits, the genetic information also seems to provide us with a definite number; DNA sequence data inform us that humans are 1.2% diverged from chimps. Case closed? In actuality, as we discussed above, the situation is more complex. Genetic information provides us with several different numbers, because that 1.2% figure only applies to nucleotide sequence substitutions of genes that code for proteins. If you include insertions and deletions, we are 4% diverged. Moreover, there is substantial variation among genes; many genes haven't changed at all between humans and chimpanzees, and others have changed many fold times the average rate. So, what exactly does that number (1.2%) mean?

In my opinion, this number based on DNA sequence data does provide us with some vital information about our relationship with chimpanzees and the other apes. The number, however, informs us more about our differences from, rather than our identity with, chimpanzees. Although much has been made of our genetic similarity, the 1.2% genetic difference in single nucleotide changes alone represents considerable genetic distance. Our genetic difference with both the common chimpanzee and the bonobo is at least ten-fold higher than the average genetic difference found between any two people within our species. Moreover, recall that the vast majority of the differences within *Homo sapiens* are among individuals of the same ethnicity and not between what have been designed as "racial groups." Given DNA sequences, a geneticist could easily distinguish humans from chimpanzees (common or bonobo) almost regardless of what genes were studied—with the exception of where persistent and strong balancing selection has maintained genetic variants since before the species diverged (chapter 5).

■ Caught Between Scylla and Charybdis?

In Greek mythology, Scylla and Charybdis are two sea monsters that live on opposite sides of a narrow strait of water. Sailors who tried to avoid Scylla would often pass disastrously close to Charybdis and vice versa. If you will permit me to mix metaphors, the chasm Huxley had mentioned regarding humans and nonhuman animals presents its own Scylla and Charybdis. One potential danger is looking at chimpanzees and the other great apes and exaggerating their differences with humans. Avoidance of that danger, however, can lead us toward a different risk: exaggerating their similarities with us.

The dangers sea voyagers faced from the two mythological sea monsters were not equal; Charybdis, but not Scylla, would create great whirlpools thrice daily by swallowing vast quantities of water and then spouting it up again. In fact, according to Homer's *Odyssey*, Odysseus was advised to steer closer to Scylla because the whirlpools of Charybdis had the ability to destroy his whole ship. Odysseus, following that advice, avoids Charybdis, though at the expense of losing six of his crew to Scylla.

So, which is the greater danger—overemphasizing the similarities between humans and chimpanzees (and bonobos) or overemphasizing the differences? I agree with Marks that questions about sameness and otherness are philosophical, not scientific, in nature. I also think that the answer to this question of the greater danger depends on what the answer will be used for. When contemplating questions of ethics, it behooves us to err on the side of our similarity with chimpanzee and other apes. A greater damage can arise if we treat them as though they are more different from us than if we treat them as more similar to us. Although I certainly do not believe great apes should be accorded the full moral standing of human beings, their phylogenetic and phenotypic similarities do impose on us the need for special consideration in how we treat them.

Yet similarity, not difference, is likely to be the Charybdis in anthropological study. I believe that overemphasis on the similarities poses a greater risk than overemphasis on the differences when contemplating our human natures. Apes are not disabled people, nor are they subhuman. *They are not human.* We are chimpanzees only in the most strict phylogenetic sense that our closest relatives are chimpanzees. We are sufficiently different from chimpanzees in so many other ways. The history of anthropology has been rife with scientists projecting our natures on those apes, and vice versa. Take the following question: Is the "warfare" seen in chimpanzee "societies" an indicator of our real nature that has been hidden beneath the cloak of culture and civilization? The first problem is that both warfare and societies need to be set in quotation marks, because the chimp versions of these abstractions need not be identical to what we humans consider warfare and society. We view both "warfare" and "society" with the biases and baggage of our own culture.

The other problem with looking to the chimps as mirrors to our nature is that we have two equally foggy mirrors that give us different views. The common chimpanzee and the bonobo, which are far closer to one other than either is to us, are worlds apart in terms of their behaviors—as different as Genghis Khan and Dr. Ruth. A pop psychologist could write a book called *Chimps Are from Mars and Bonobos Are from Venus.* We are just as close in genetic terms to the bonobo as we are the chimp. We share just as many genes with each species. But just as neither Venus nor Mars is planet Earth, neither bonobos nor chimps are human.

What Are the Genetic Differences

That Made Us Human?

Blasé, matter-of-fact, encumbered by the infant (who, face to
her chest, clutches her fur), she carefully positions the hard-
shelled fruit on the log and smashes it open—using a stone
tool procured for the purpose. Hammer and anvil. No light
bulb goes off above her head. There's no chin to fist, no hint of
insight struggling to emerge, no moment of revelation, no
strains from Also Sprach Zarathustra. *It's just another routine,*
humdrum thing that chimps do. Only humans, who know
where tools can lead, find it remarkable.

—Carl Sagan and Ann Druyan (Sagan and Druyan,
1992, p. 391)

∎ The Needles in the Haystack

As we saw in the previous chapter, the extent to which we are different or
similar to chimpanzees and other apes—or for that matter other animals in
general—depends upon one's point of view and the types of questions that
one asks. Although genetics can inform debates revolving around the same-
ness-otherness dichotomy, such questions are philosophical in nature, as
Jonathan Marks stated.

Regardless of how close to or how far apart we consider ourselves from
chimps, we certainly differ substantially from our relatives among the apes
with respect to brain size and capacity for language. The vocabulary of
four-year-old human children outstrips that of apes—even those apes that
researchers assiduously trained in language. No other animal has composed
sonnets or, on the flip side, a *Mein Kampf.* Although several of the great apes
use tools, no other animal has built anything resembling a go-cart, much less
a microprocessor.

As exemplified by the quote from Sagan and Druyan, observations of other
species of apes often leads to contemplation of our origins. What made us
different from them? Is it a case of "there but for the grace of different selective pressures..."? Can we pinpoint the genetic changes that made us
human? The answer to the last question depends upon what one means
by human. Such a question, laden with philosophical overtones, may never be
answered. Evolutionary geneticists are, however, becoming better able to

detect the changes that occurred on our lineage and which have been driven by positive selection. They also are increasingly adept at determining the genetic changes that have affected our brain size and linguistic capacity during our evolution.

A large part of the problem in detecting such changes is that our genome is so vast in size. From the perspective of the genome as a whole, the genetic differences between humans and chimps are overwhelming. If one examines just single nucleotide changes, 40 million nucleotides differ between humans and chimps! (Even more genetic divergence between these species arises from small deletions and insertions of DNA.) We have no reason to assume that one lineage changed considerably more than the other; thus, roughly half of these changes occurred in each lineage, so that means approximately 20 million nucleotide substitutions occurred on one lineage since we split from chimps six million years ago. If one were to devote 12 hours a day to just counting the single nucleotide differences between humans and chimpanzees, this task would take 16 months (assuming one could count one nucleotide difference per second). The vastness of the genome means that humans and chimps differ at a huge number of sites, even though only a small proportion of our genome has changed.

Out of these tens of millions of nucleotide differences, what were the ones responsible for the "important" differences between humans and chimpanzees? The DNA differences that are responsible for changes in brain size, changes in skull shape, use of language, gait, the ability to think in abstractions, and other traits of interest are but a vanishingly small subset of the total change that has accumulated. Even if these differences were due to as many as 10,000 nucleotides that changed along our lineage, that would still be much less than one thousandth of the total changes.

Evolutionary geneticists use two approaches to find the genes involved in the important differences between humans and nonhuman apes. The first begins by considering genes with known effects on these traits of interest. Through examination of the evolutionary history of each of these particular genes, researchers can determine whether positive selection has driven changes in their sequence in our lineage. As you might suppose, such an approach is laborious; it requires intensive study of one gene at a time. Possessing the genomic sequences of humans and chimpanzees allows evolutionary geneticists to take a second approach in which they do their detective work at the "wholesale," rather than the "retail," level. In this second and complementary approach, geneticists start with a culled list of genes taken from the whole (or nearly whole) genome; these culled genes all show the footprints of the action of positive selection along the human lineage. Geneticists then can match these genes with functions.

We begin with an example of the first approach and an unusual genetic pedigree.

■ A Complex Language Defect with a Simple Genetic Basis

The discovery aspect of science at times seems to work in mysterious ways, with serendipity playing an important role. For instance, I'm sure that the researchers studying a language disorder that was passed on as a simple Mendelian trait in one family never dreamed that their work would help in discoveries about the evolution of linguistic abilities in humans, and yet that appears to have happened.

The story begins with the finding of a language and motor disorder passed on for three generations in one British family.[1] This family (known as the KE pedigree) and their disorder came to the attention of Faraneh Vargha-Khadem, an Iranian-born neurologist at the Institute of Child Health in London, and a group at Oxford University led by Anthony Monaco and Simon Fisher. These researchers noted that the disorder appeared to be inherited like a textbook case of a single gene, where the variant that caused the affliction was dominant. Contrary to popular belief, most disorders that are prevalent in the population, even those that "run in families," do not exhibit simple patterns of inheritance (e.g., one gene with a variant that is either recessive or dominant). Instead of behaving like the traits Mendel studied in peas, most traits that affect the length and quality of human life have complicated patterns of inheritance. For most common disorders—coronary heart disease, Alzheimer's, schizophrenia, diabetes, and several others—no one-to-one relationship exists between a genetic variant and the disease. Possessing certain genetic variants can affect one's propensity for acquiring a disorder and sometimes the severity of the disease, but whether one acquires a disease is influenced by a complex relationship of the actions of several genes, as well as several environmental factors. For the most part, human geneticists have nearly exhausted unraveling the causes of the "easy" cases, such as sickle cell anemia and phenylketonuria (PKU), which do follow simple patterns of inheritance; most remaining traits display more complex patterns. So the defect found in the KE family is unusual in the simplicity of its genetic basis. The apparent simplicity of the genetic basis gave the researchers hope that they would be able to quickly pin down the gene or genes that caused the defect. Before discussing the genetic details of this disorder, let's take a closer look at its manifestations.

This language defect found in the KE family, although simple in its genetic basis, is complex in the suite of characters it affects. Individuals with the disorder struggle with articulating words because they lack proficiency in performing the correct movements of the mouth and face (particularly of the upper lip) needed to produce intelligible speech. The troubles they face are not just with making the correct muscular movements, but also with producing these movements in the proper sequence. Affected individuals also are deficient in understanding grammar and in processing language; in particular, they have trouble breaking words down into individual sounds. In addition to severe defects in verbal intelligence, these individuals appear to have more

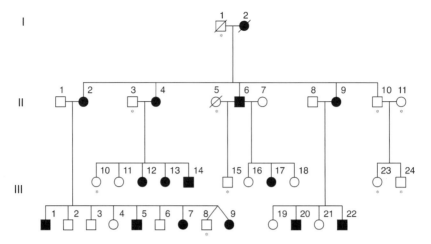

Figure 9.1

The KE family pedigree. Circles and squares represent females and males, respectively. Shaded individuals are affected by the severe language disorder, whereas unshaded individuals are not. Figure courtesy of Simon Fisher.

subtle impairments in nonverbal intelligence, as indicated by the nonverbal IQs of most of the affected individuals tested being in the low-average (80–89) range.

One diagnostic trait of this condition is that afflicted individuals score far worse than nonafflicted individuals on tests involving repeating words (or nonword sounds) back to the instructor immediately after the instructor speaks them. The range of scores of affected individuals falls outside the range of that of nonaffected individuals. Affected individuals had particular difficulty in repeating words with more than one syllable. This lack of facility with multisyllabic words is further evidence that the defect involves difficulty in coordination and in making the correct movements in the proper sequence.

The syndrome seen in the KE family has some similarities to patients with a well-known disorder: adult-onset Broca's aphasia. Right around the time Darwin published *The Origin of Species*, a French physician named Pierre Paul Broca was examining people who had had strokes or other brain damage that limited their ability express words. Their language would be halting, often reduced to simple monosyllabic words. Sound familiar? Broca found that this inability to speak (technically called aphasia) in these patients arose due to defects in a region of their frontal lobes on the left sides of their brains. The brain region affected in these patients is now known as Broca's area.

The brains of affected individuals are also different. Positron emission tomography (PET)[2] scans testing the brain activities of subjects when they were engaged in tasks involving word repetition revealed clear functional brain abnormalities. It is interesting that in the affected subjects, certain brain regions were overactive as compared with controls, but other regions were underactive. Even more intriguing, one of the overactive brain regions in affected individuals was Broca's area, the paradigmatic "language center" of the brain. It is possible that hyperactivity in Broca's area reflects overcompensation of affected individuals struggling to repeat words correctly. Structural magnetic resonance (MRI) tests revealed no clear-cut pathologies, but statistically significant differences exist between affected and control individuals. For instance, affected individuals tend to have less gray matter in Broca's area than do control individuals. Moreover, functional MRI[3] tests, in contrast with the PET scans, show less activity in Broca's area in affected individuals.

Through molecular genetic analysis, researchers mapped the disorder in the KE family to a defect in a single gene that maps to chromosome 7. In addition, the researchers came across a case study of an individual unrelated to the KE family but who had a similar language syndrome. It turned out that this individual had a translocation involving the region of chromosome 7 corresponding to chromosomal position to which the trait in the KE family mapped.

Subsequent work in characterizing the gene involved in this syndrome revealed that the gene was part of the Forkhead-wing-helix gene family (FOX), and was thus designated *FOXP2*. FOX genes play a number of roles by regulating other genes. The proteins of these FOX genes contain what is known as a "forkhead box"—a region of 80 to 100 amino acids that binds to DNA.

The syndrome seen in the KE family arises from a single nucleotide change (G to A) that causes an amino acid change in position 553 of the protein. Notably, this nucleotide change is present in all affected individuals of the KE family, and all unaffected individuals from that family lack the nucleotide change. This mutation has not been detected in the general population, thereby implying that it is either absent or very rare. The affliction is dominant because two functional copies of the gene are needed for proper language, and hence the defect shows up when there is only one functional copy.

FOXP2 interacts with a number of downstream genes in regulatory pathways. By binding to regulatory sites of these genes, *FOXP2* reduces the rate at which the genes are transcribed into RNA. In humans and probably other vertebrates, *FOXP2* itself is expressed in the brain and several other organs during several stages of fetal development. Although its exact role in gene regulation is still being deciphered, biologists know that *FOXP2* interacts with several related proteins of the forkhead family. Some scientists speculate that speech development and language abilities require having the proper balance of expression of these forkhead genes.

■ FOXP2 and Human Evolution

One of the striking aspects of the evolution of *FOXP2* is its evolutionary conservation. Within mammals, the FOXP2 protein is one of those evolving at among the slowest rates. In humans and mice alike this protein contains a total of 715 amino acids, but only three amino acid sites changed between humans and mice, which are about 80 million years divergent. This conservation isn't limited to just mammals; the human and bird (zebra finch) FOXP2 proteins are only 2% different.[4] This means that in the 300 million years of evolution since birds and mammals split, the sequence of this protein has diverged only 1% in each lineage.

Svante Pääbo's group has investigated the evolution of FOXP2 in humans and other apes. They found that two of the three amino acid differences between humans and mice occurred on the human lineage after humans split from chimpanzees.[5] Note that neither of these changes is the mutation that was found in the KE family. One change along the human lineage could be happenstance; the presence of two changes along the human lineage just by chance is improbable, especially when one considers that the split of humans from chimpanzees is less than one-tenth the split between the human/chimp and the mouse lineages. One could argue that such changes are just due to the relaxation of constraint, but as we will see, further evidence rules out this hypothesis of the relaxation of negative selection.

In humans, FOXP2 has more rare genetic variants than would be expected under a neutral model in which the population size is constant and individuals mate at random. Population growth would lead to an excess of rare genetic variants, but this effect should be the same across the whole genome and not limited to a single gene. Although most genes in the human genome do show a small increase in these rare variants, the deviation seen at FOXP2 is unusually large. This suggests that something in addition to the increase in population size is occurring at FOXP2. That something else is likely to be a recent selective sweep.

Pääbo's group estimates that this positively selected variant was fixed sometime in the last 120,000 years. The estimate is based on a model that assumes random mating and no population growth. With the inclusion of population growth, the window of the estimate can be pushed back in time somewhat, but only to not more than 200,000 years ago. Even with this consideration, it seems likely that this gene was fixed in the human lineage after the split of humans and Neanderthals.

Several popular science accounts of this finding have called FOXP2 "the language gene" or "the speech gene." For instance, cnn.com's account on August 15, 2002, of the finding by Pääbo's group bore the subheading, "Apes lack gene for speech."[6] This statement is problematic for several reasons. First, FOXP2 is not the only gene involved in speech or language; we do not even know whether it is the most important gene affecting speech that has changed on the lineage. It's just the first gene of its kind to be discovered. Second, apes don't lack the gene; they have a perfectly functioning FOXP2 gene; it just differs somewhat from our human version of that gene. Finally, it isn't clear whether the change in FOXP2 that occurred in our ancestors had anything to do with speech per se. Positive selection certainly operated on the gene, but for what, we just don't know.

As anthropologists, cognitive scientists, linguists, and others have observed, language is not a unitary quantity. As Ray Jackendoff puts it, we shouldn't be asking whether apes or "hominids" had language, but rather what components of language they possessed.[7] We know that trained apes, especially chimpanzees and bonobos, have shown considerable success at learning some aspects of language but fail to learn others. We also know from our own experiences that developing children do not acquire language in one step, or even two steps. Why should the evolutionary development of language come about in a single leap? This isn't to say, to paraphrase Ernst Haeckel, that the ontogeny of language recapitulates its phylogeny; there is no reason to suspect that the steps in which babies learn language will necessarily match the steps by which human ancestors evolved language. Still, it is likely that in the course of our evolution, we acquired language in a stepwise manner.

■ Shrunken Brains

Aside from our enhanced speech and linguistic abilities, the other most striking difference between humans and other apes is the size of our respective brains. Human brains are three to four times larger than those of our closest living relatives; the size difference is the result of several episodes of expansion occurring in the human lineage. Most notably, brain size showed a fairly steady size increase from just under two million years ago until about 500,000 years ago. In the last half a million years, brain size remained at a plateau; indeed, Neanderthal brains were slightly larger than our own.

Microcephaly—literally "small head"—can arise from a number of environmental insults during development (including exposure to harmful chemicals, poor nutrition, or maternal diabetes). This condition can also arise from the loss of function of any one of at least six genes. Although people with genetic microcephaly have drastically smaller brains (typically one-third the normal size at adulthood), as well as significant to severe mental retardation, their brains are relatively normal in overall structure. This suggests that the genes, whose lack of function results in microcephaly, play some role in regulating brain size. Could positive selection have acted on these genes to lead to larger brains in humans?

Bruce Lahn and his colleagues at the Howard Hughes Medical Institute at the University of Chicago have been investigating the evolution of genes; they found that positive selection has operated on at least two of the six known "microcephaly" genes: *ASPM* and *Microcephalin*. The stories that these two genes tell are similar: positive selection has been acting on these genes for a long time in primates and has acted in the very recent past of humans. We'll start with *Microcephalin*.

In primates, *Microcephalin* exhibits a high ratio of replacement site changes to silent site changes.[8] Led by Lahn's graduate student Patrick Evans, these researchers found that throughout the last 20 million years of primate evolution, the replacement to silent changes ratio has been particularly high along the lineage that led to humans as compared with sister lineages that did not lead to humans. Recall that to definitively ascribe positive selection as the cause of a high replacement to silent ratio, the ratio has to be statistically larger than one. Because this ratio turns out not to be consistently statistically larger than one along the human lineage, the formal possibility remains that the increased number of replacement site changes are due to the relaxation of constraints in the human lineage, and not the action of positive selection.

To rule out the lifting of constraints hypothesis, Lahn's group also collected polymorphism data from human populations, which enabled them to perform McDonald-Kreitman tests. Recall that in the absence of positive selection, the ratio of replacement to silent site polymorphisms within species should be statistically indistinguishable from the ratio of replacement to silent changes between species. In actuality, these ratios were very different; positive selection has driven an estimated excess of 45 amino acid changes at this gene in the last 20 million years of primate evolution along the branch that led to humans.

The polymorphism data that Lahn's lab collected for the previous study led to another conclusion: one haplotype of human *Microcephalin* was found to be at a much higher frequency than that of the others.[9] By looking at the chimpanzee gene, the researchers could infer that this common haplotype differed from what had been present in the common ancestor of chimps and humans. Evolutionary biologists call such a change a derived one. This derived haplotype is nearly fixed in populations indigenous to the Americas, is very high in frequency throughout most of Europe (including the Middle East) and Asia as well as in New Guinea, but is low to moderate in frequency in sub-Sahara Africa.

Lahn and colleagues then calculated that the probability of observing a derived haplotype at such a high frequency simply by chance was extraordinarily low. Their calculations also showed that this haplotype went from being present in only one copy (as a new mutation) to a high frequency in a relatively short time—a period too short for the frequency rise to be due solely to random genetic drift. Thus, the authors conclude, positive selection must have operated on this variant. The estimated age of this variant was 37,000 years, with a range between 14,000 and 60,000 years. This estimate is right within the range of the "Great Leap Forward," the cultural explosion that Jared Diamond proposed (see Interlude II). The variant's high frequency in New Guinea, which was settled around 40,000 years ago, and the Americas would be difficult to explain if the mutation arose toward the more recent end of the range. Other data also support the recent age of the haplotype, as well as the action of positive selection.

The *ASPM* gene tells a similar tale. Based on the McDonald-Kreitman test, this gene also has had an excess of amino-acid changing substitutions along the human lineage. In fact, the test suggests that 15 of the 19 sites that change the amino acid between humans and our common ancestor with chimps have been adaptive.[10] This would imply, given a six-million-year divergence time, that there has been an amino acid change that has been driven by positive selection once every 400,000 years on average. Other tests suggest that adaptive evolution has been operating on this gene on our lineage for almost 20 million years.

One haplotype of *ASPM* in humans is also much more common than all of the others and is also a derived state.[11] It is at highest frequency in European and Middle Eastern populations, relatively high in Central Asia, and rare to very rare elsewhere. This haplotype appears to be even younger than the one found in *Microcephalin*, with an estimated age of less than 6,000 years (the range of this estimate is between 500 years and 14,000 years). By population genetic standards, that's overnight. By paleontological standards, that's a blink of an eye. By the same reasoning used with the *Microcephalin* haplotype, the researchers conclude that positive selection has driven this variant to high frequency.

So what does this mean? One phrase that is often used in scientific papers is "it's tempting to speculate." I usually loathe this phrase because authors using it most often really mean "we speculate that, but we don't want to come out and say that we speculate it." In this case, the phrase is warranted. *It is tempting to speculate about the implications of these findings.* Does it mean that humans with those haplotypes have bigger brains and/or are smarter than those who lack those haplotypes? In addition, the nonrandom distribution of these haplotypes across the globe poses huge ramifications.

Let's resist the urge to speculate wildly, lest we lead ourselves into temptation. Let's consider what we do know. The inference is that recent positive selection drove the increase in frequency of these genetic variants. But we don't know much more. Human brain size is not increasing and hasn't increased for hundreds of thousands of years. In fact, it's possible that our brains have decreased somewhat in size during that time; Neanderthal brains were bigger, on average, than those we possess. Could it be that selective pressure has operated to reduce our brain size, and that these haplotypes at *ASPM* and *Microcephalin* actually lead to smaller brains? We don't know.

Moreover, as Gould has pointed out in *The Mismeasure of Man*, no solid evidence supports a large correlation between brain size and intelligence within the normal range of brain size.[12] Let's be careful out there.

■ Lose a Gene, Grow a Brain?

Although most of this change is likely to be due to changes in genes, one intriguing hypothesis is that the growth of our brain has been assisted by the fortuitous loss of function of a gene. How could losing a gene help make brains larger?

Sialic acids are a major component of chemicals that coat the surfaces of cells. These chemicals play diverse roles in a large number of functions, including immune defense, cell signaling, differentiation of cell types, and the transport of ions. Unlike all other apes (and probably all other mammals), humans lack one of these sialic acids, N-glycolylneuramic acid (Neu5Gc). Humans don't have this chemical because an *Alu* insertion inactivates the gene that produces the enzyme (CMAH) that converts the precursor Neu-5Ac to Neu-5Gc. In contrast, Neu-5Gc coats the cells of other apes, including both species of chimps, because the enzyme CMAH is functional in these species. Because we lack CMAH, we humans also have an increase in the content of Neu5Ac, the precursor sialic acid.

Brain growth ceases soon after birth in most primates. In humans, however, brain growth continues on. One hypothesis is that brain growth is inhibited by the presence of Neu5Gc. In nonhuman mammals, the expression of the *CMAH* gene is reduced in the brain, leading to less CMAH enzyme. The consequence of this down-regulation is that in nonhuman mammals, the surfaces of cells in the brain have less (but still have some) NeuGc as compared with cells elsewhere.

Two methods were used to date when the *Alu* insertion inactivated the *CMAH* gene.[13] The results from both of these methods support the insertion and inactivation of *CMAH* occurring about 2.7 million years ago, shortly before the rapid expansion of our brains.

So did Neanderthals have a functional version of this gene? Because this gene is nuclear and not from the mitochondria, it would be exceedingly difficult to obtain enough DNA in order to sequence it. It is not necessary to sequence their DNA to find out whether Neanderthals had the functional CMAH enzyme; researchers can use another method—by looking at the Neanderthal's sialic acids. Neu5Gc was not detected in Neanderthals, but researchers did find Neu5Ac, its precursor, in Neanderthals.[14] This is strong evidence that Neanderthals also have an inactive *CMAH* gene, as we would expect, given what we know of the timing of the inactivation.

The weakest link in this chain of argument about the inactivation leading to increased brain size is whether Neu5Gc indeed inhibits brain growth. The inactivation of the enzyme dating to around the same time that our brain size exploded is intriguing. But remember the old adage, "post hoc, propter hoc." Just because event X occurs before event Y doesn't mean that X caused Y.

The loss of Neu5Gc and subsequent increase in Neu5Ac in humans appears to have another consequence: changing our susceptibility to malarial parasites.[15] *Plasmodium falciparum*, the major malarial parasite in humans, does not cause severe infection in chimpanzees. Conversely, *Plasmodium reichenowi* causes malaria in chimpanzees but apparently not in humans. The protein from *Plasmodium falciparum* involved in binding to red blood cells prefers to bind to cells with Neu5Ac; it is actually inhibited by Neu5Gc. But the binding protein from *Plasmodium reichenowi* prefers Neu5Gc. Manipulations of the sialic acid content of human cells in cell cultures show that it is indeed the composition of sialic acid that alters the binding preferences of the proteins.

■ Taking It to the Genome

Instead of taking the "candidate gene" approach—starting with genes of known function and then examining the molecular evolution of the genes—one could start with all or most of the genes, find those with the signatures of selection, and then look at the functions of those genes. Because this genomic approach is the reverse of the standard genetic approach, it is sometimes referred to as reverse genetics.

Prior to the publication of the chimpanzee genome by the publicly funded consortium, the private company Celera had sequenced the genome of this species. Cornell University evolutionary geneticist Andrew Clark and some of his colleagues collaborated with scientists from Celera in the first genome-wide analysis with the aim of finding genes whose evolution was accelerated along the human lineage by positive selection.

The approach taken by the Cornell and Celera researchers allowed for variation in the ratio of replacement to silent sites across the gene; this way, they could detect positive selection even if the overall gene did not have an excess of replacement changes, if regions of the gene did. In their first report, they also used the mouse sequence as an outgroup to look at changes that occurred only along the human lineage. If humans have a G at a site, chimps have C, and also mice have C, one can infer that the change was in the human lineage. Through this method, Clark and his colleagues found that hundreds of genes showed signatures of the action of positive selection—namely, more than the expected number of replacement site changes in at least some regions of the gene.[16]

Clark and his colleagues could infer that positive selection had operated on *FOXP2*, the gene involved in language, as well as many other "brain" genes. Although genes that played a role in the brain (particularly neurogenesis, the generation of new neurons in development) were disproportionately overrepresented in the genes that were accelerated by positive selection, this was not the only overrepresented category of genes. Genes involved in the breakdown of amino acids and genes involved in hearing were also highly represented on the list. Amino acid breakdown could be under selection due to dietary changes that occurred in our evolution.

The "hearing genes" category is an interesting case. One of the hearing genes, *alpha tectorin*, is involved in the development of the tectonal membrane in the inner ear. Mutations in this gene can impair hearing, particularly at frequencies in the range at which words are often spoken. This gene may have been under selection because the development of language also places demands on hearing and processing of particular sounds.

In 2005, 18 months after their first paper, the Cornell-Celera team published a second.[17] In this study, they had a larger number of genes and they did not use an outgroup to determine the direction of the change between humans and chimps. Instead of investigating the changes that happened just on the human lineage, they looked at all changes between humans and chimpanzees. Thus, the first study concentrated on genes that had been influenced by positive selection in just the human branch, whereas the second study examined positive selection acting during the histories of both humans and chimpanzees (including bonobos).

Immune-defense genes were overrepresented on this second list of genes under the influence of positive selection. This finding shouldn't come as a huge surprise, because pathogens collectively have been and remain a major contributor to mortality. Moreover, because "arms races" are often involved in defenses to pathogens, we would expect genes that influence our responses to them to have undergone multiple rounds of selection. Immune system genes and genes involved in reproduction are usually those that differ between animal species. It's a bit depressing from our vanity, but we're just not as unusual as we like to think we are.

Evolutionary biologists have long known that arms races are also a likely consequence of the action of sexual selection. Competition between males for access to females, however, is not just limited to mating; sexual selection can continue even after copulation, in the form of sperm competition, if females mate more than once. During the 1990s, evolutionary biologists accumulated considerable data that support the contention that this cryptic form of sexual selection is pervasive across the animal kingdom.[18] Humans and other apes are not likely to be exceptions; and in fact, sperm competition seems to have been a potent force in our evolution. Consistent with this thesis, many genes involved in spermatogenesis show accelerated evolution, which was probably driven by repeated rounds of positive *sexual* selection.

Interestingly, in this second set, there was no tendency for those genes that expressed most in the brain to show the hallmarks of positive selection. Indeed, these are among the most conserved genes! The differences between the rates of evolution in the "brain genes" between the first and the second publications could be explained by the fact that the first paper considered changes that occurred in our lineage after it separated from chimpanzees, whereas the second examined overall differences between humans and chimpanzees.

In this second report, the researchers picked the 50 genes with the strongest statistical signal for an accelerated rate of amino-acid substitutions, presumably due to the action of positive selection. Note that these are not necessarily the genes that were subject to the most positive selection; these are the genes for which the evidence is strongest that positive selection operated on them. Many of the "top 50" genes are involved in the defense against pathogens, and four are involved in olfaction.

A considerable fraction of the top 50 genes appear to have roles that could influence the formation of cancer; some are tumor suppression genes and some are involved in the control of the cell cycle. One of the genes with the strongest signal is *TSARG1*, which is involved in programmed cell death[19] during spermatogenesis. In programmed cell death, particular cells shut down in response to chemicals and/or signals from other cells; they seem to "commit suicide." In the last few decades, biologists have seen that this programmed cell death is a necessary and important process in the development and function of organisms. Cellular tumors are often the result of improper functioning of programmed cell death.

As the Cornell authors note, the elimination of germ cells posed by this programmed cell death can set up what evolutionary biologists call a genomic conflict. Certainly, the "cheater" individual germ cells that can avoid programmed cell death will have an advantage over cells that do not. But the avoidance of programmed cell death could be deleterious to the organism as a whole. Suppose a new mutation with a slightly greater chance of avoiding programmed cell death appears in the population. This variant will thus be under positive selection, which could be rather strong. If this variant increases in frequency and increasingly more cells escape the action of the grim reaper, the functioning of the organism as a whole is compromised. This sets up the conditions for other mutations to evolve that repress the ability of "cheater" cells to escape programmed cell death. There is even the possibility of an "arms race" through multiple rounds of cells becoming better able at escaping programmed cell death countered by more effective "policing" by other cells. Such a scenario for why *TSARG1* has seen so much positive selection is, of course, speculative, but not unreasonable. It also provides a reason to study *TSARG1* more intensely.

The authors also took these "top 50" genes and sequenced them in 20 Caucasians and 19 African Americans. In general, they found more genetic variants that are derived and at high frequencies than would be expected under a neutral model of just mutation, genetic drift, and negative selection. This is further support for these genes being under positive selection. One of the top 50 genes, *OR5I1*, which is involved in olfaction, showed a different pattern in the polymorphism study; it had many more polymorphisms segregating within the populations than expected. This result suggests that what had initially been posited to be positive selection may actually stem from balancing selection.

So where does this leave us? It is clear that positive selection repeatedly has left its impact on our evolutionary history. The genome-wide studies demonstrate that positive selection—natural and sexual—has driven the evolution of a large number of genes since we diverged from chimps and bonobos. The overall number of genes exceeds a thousand, and could be much more. Although some "brain genes" have been subject to repeated selection, many more selection-driven changes are in genes with seemingly more mundane functions, like pathogen-protection and spermatogenesis.

Has the evolution of humans been typical of large mammals? We do not yet have a comparable pair of large vertebrate species to make the appropriate comparisons. Given the accelerating pace at which genomic studies are performed, however, such studies will be likely in the near future.

The implications of research into the changes that have occurred along the human lineage may extend beyond the realm of evolutionary anthropology and into applied medicine. For instance, Lahn's group has argued that such evolutionary studies can also be used to spur on and provide guidance to future studies of medical genetics.[20] Specifically, consider genes that (A) show the hallmarks of accelerated evolution via positive selection in the human lineage and (B) are expressed in brains (or have some known brain function). It could be worthwhile to look for mutations in these genes in individuals that have microcephaly (or other structural brain abnormalities) with unknown genetic cause.

Clicks, Genes, and Languages

If we possessed a perfect pedigree of mankind, a genealogical relationship of the races of man would afford the best classification of the various languages now spoken throughout the world; and if all extinct languages, and all intermediate and slowly changing dialects, were to be included, such an arrangement would be the only possible one.

—Charles Darwin (Darwin in Cavalli-Sforza and Cavalli-Sforza, 1995, p. 167)

∎ Click Languages

At the start of the 1980 movie *The Gods Must Be Crazy*, a westerner throws a glass bottle from a plane, and that bottle lands in the Kalahari Desert in southern Africa. An indigenous tribe of hunter-gatherers discovers the bottle. Enchanted by the glass bottle, they believe it is a gift from the gods. Unfortunately, there is only one bottle, and it cannot be shared among this egalitarian tribe. The tribe, seeing the jealousy and strife brought about the bottle, comes to the conclusion that the bottle is evil, which must be cast out. They send off an envoy, Xixo, to return the bottle to the gods. One other notable feature about this tribe is that their speech is punctuated with clicking sounds.

Although numerous liberties were taken in this movie, it is grounded in fact. Egalitarian bands of hunter-gatherers have existed in the Kalahari Desert for millennia. Until the last few decades of the twentieth century, most of these groups lived in complete or near complete isolation from each other. In several of these tribes, the repertoire of sounds in their language includes clicks. One of these, the !Kung (also known as the Ju/'hoansi), was the model for the tribe in the movie. The "!," as in !Kung, signifies a click, which is created by agile tongue movements coupled with a rapid inward movement of air.

Approximately 120,000 people today speak the 30 or so extant click languages, which form the Khoisan language "superfamily"; these languages are currently found only in sub-Sahara Africa.[1] Clicks are also found in the neighboring Bantu languages, including Xhosa, which is spoken by eight million people and is one of the official languages of South Africa. The most prominent speaker of Xhosa is Nelson Mandela, the anti-apartheid activist and the first democratically elected president of South Africa. He was born Rolihlahla Mandela and comes from the Thembu Xhosa family. The clicks that are in the Bantu languages were probably picked up as the Bantus spread southward into the lands occupied by Khoisan speakers two or three millennia ago. Clicks are not used in languages outside the Bantu and the Khoisan languages.[2]

As we will see later in the chapter, genetic tools have been used to determine the age of these Khoisan click languages. Despite their rarity, these languages appear to be ancient—perhaps tens of thousands of years old. We will also discuss the possible functional roles that thee click sounds may play in the daily activities of these hunters. But first, let's turn to general principles of language evolution.

■ Early Language Evolution Studies

Literate medieval speakers of Spanish, French, and Italian knew that their languages had descended with modification from Latin because they had a historical record of these changes. An early example of the recognition of language change occurred at the signings of the Oaths of Strasbourg in 842. Here, Charles the Bald and Louis the German—the two younger grandsons of Charlemagne and rulers of western and eastern Frankish kingdoms, respectively—pledged their allegiance to each other and their older brother Lothair, the Holy Roman Emperor. Charles and Louis realized that the local "Latin" vernacular in Charles' land had diverged so far from "Classical Latin" that the commoners could not understand classical Latin. Therefore, they took the Oaths of Strasbourg in the vernacular of the other's land: Louis in a proto-French and Charles in the language spoken by the eastern Franks of the time.[3]

The Germanic languages, including forms of English, were not derived from Latin, but medieval scholars had observed sufficient commonality between the Romance and the Germanic languages that the scholars knew that the languages probably shared a common ancestor. Similar patterns among most European languages had also been observed in medieval times.

Recent studies linking genetic and language studies confirm that Slavic (which includes Russian and Polish, as well as the Baltic languages), Romance, and Germanic language families are each true language families. Within each family, all languages share the same common ancestor, and all languages that share the common ancestor are included within the language family. There is also strong support that all three of these language families are very close to one another, thus constituting a language superfamily that also includes several other families: Baltic, Celtic, Hellenic, Iranian, Indic, Armenian, Albanian, and Tocharian. Among the Germanic, Romance, and Slavic language families, evidence points to the Germanic and Romance families being the most closely related (with the Slavic family as the out-group), but the support for that finding is less strong than it is for the three families being each another's closest relatives.[4]

The first real breakthrough regarding the evolution of languages came from the British jurist and ancient India expert, William Jones, during the latter part of the eighteenth century. Jones put forth the argument that ancient Sanskrit, ancient Latin, ancient Greek, and the other European languages were derived from a common ancestor. He stated:

> The *Sanscrit* language, whatever be its antiquity, is of a wonderful structure; more perfect than the *Greek*, more copious than the *Latin*, and more exquisitely refined than either, yet bearing to both of them a stronger affinity, both in the roots of verbs and the forms of grammar, than could possibly have been produced by accident; so strong indeed, that no philologer could examine them all three, without believing them to have sprung from some common source, which, perhaps, no longer exists.[5]

Jones noted that the common ancestor of these languages might have gone extinct, but he hoped that it could be reconstructed. The scope of language studies was thus widely expanded; Jones's studies established the Indo-European superfamily of languages and allowed for the possibility that other non-European languages would also be connected in this language family tree.

Fast forward to the early 1860s, just after Darwin published his theories of evolution. During this time, the German linguist August Schleicher published perhaps the first account of the similarities of the processes of biological and language evolution. His friend, the eminent biologist Ernst Haeckel, had been pestering Schleicher to read Darwin's *Origin*, which had just been translated into German. After he finally got around to reading Darwin's work, Schleicher noted how Darwin's principles of variation, inheritance, and selection were operating in his garden, where weeding was the agent of selection. Schleicher went on to state that "the rules now, which Darwin lays down with regard to the species of animals and plants, are equally applicable to the organisms of languages."[6]

Studies of language evolution have had a large head start over studies of biological evolution for several reasons. First, languages evolve much faster than do most of the large, multicellular organisms. (Microbes certainly can evolve very quickly, but essentially nothing was known about microbes during the eighteenth and even the nineteenth centuries). Linguists of the eighteenth century, in contrast to biologists, could easily see the evolution of their object of study. Moreover, language evolution—although counter to a strictly literal interpretation of the Biblical Tower of Babel—does not threaten human uniqueness as long as humanity is the only form of life that possesses language.

■ The Origin of Language versus Language Evolution

The origin of language, as distinct from its evolution, was a different story. At the same time that Schleicher was constructing a Darwinian explanation for the evolution of languages, the Linguistics Society of Paris banned discussion regarding the origin of language, mainly because it was far too speculative a subject for their scientific meetings.

In a 2005 book on language origins, Robbins Burling, a University of Michigan linguist who has written extensively on the subject of how language evolved, said the following about the ban: "The ban is often cited as a sorry example of intellectual censorship, but anyone who has read widely in the literature on language origins cannot escape a sneaking sympathy with the Paris linguists. Reams of nonsense have been written on the subject."[7]

Burling speaks from many decades of experience as a linguist. As an evolutionary geneticist coming to the literature on language origins from the outside, I feel a similar exasperation. Numerous and varied explanations have been put forth to explain how humans acquired language—many of which date back to the time of the Paris ban. Some of the more fanciful explanations have been given nicknames, mainly to the effect of dismissal by ridicule. The scenario by which language evolved in humans to assist the coordination of working together (as on the pre-historic equivalent of a loading dock) has been nicknamed the "yo-heave-ho" model. There's the "bow-wow" model in which language originated as imitations of animal cries. In the "poo-poo" model, language started from emotional interjections.

During the twentieth century, and particularly its last few decades, discussion of language origins has become respectable and even fashionable. One major problem remains, however; most models about language origins do not readily lend themselves to the formation of testable hypotheses, or rigorous testing of any sort. What data will allow us to conclude that one model or another best explains how language arose?

As we discussed in the previous chapter, many researchers who study language origins believe that language arose not in one fell swoop but in a step-wise progression. The step-wise nature of language evolution—though in my opinion, more realistic—complicates things. It is quite likely that the various aspects of language could have evolved for different purposes, and thus no unitary explanation exists for the origin of language.

For the remainder of this chapter, we will not consider the origin of language, but rather how it has evolved in the time since fully formed languages have been in existence. Just as biologists know far more about the evolution of life on this planet than they do regarding how life originated, linguists are on far surer footing regarding language evolution than they are with respect to exactly how language originated.

■ Language Evolution versus Biological Evolution

Although some clear parallels can be found between language evolution and biological evolution, important differences are also apparent. As previously discussed, language evolves much faster than do most genes. Over the course of just 1,000 years, two languages separated from a common ancestor will have diverged from each other by an average of 26% of their words.[8] Similar to genetic evolution, wherein considerable variation exists among the rates at which different classes of genes evolve, the rate at which various types of words change also varies. Slowly evolving words usually are the ones most commonly used; more elaborate, less utilitarian words are freer to change.

Another critical difference between language evolution and genetic evolution is the mode of transmission of the elements. Nearly all multicellular organisms inherit their genes from one or both of their parents. Genes are transmitted vertically from parent to offspring. In contrast, elements of language are transmitted to individuals, not just from their parents but also by various other members of society. A child learns words and their use from teachers and peers, as well as from her parents. In recent years, various media (television, radio, newspapers, the Internet) have also played a role in shaping the language of children. For those who are parents, think of the words that your child picked up that you wish he or she hadn't. Even as adults, other people influence elements of our language. New words and phrases—for instance, "Internet," "genome," and "War on Terror"—are continually added to the lexicon, and other words change their meaning. Unless we invented the new words ourselves, we necessarily learn these words and phrases from others. All modes of transmission that are not parent-to-offspring are called horizontal transmission.

Horizontal transmission can also occur between different languages. The word "tsunami" is of Japanese origin and literally means harbor wave (*tsu* = "harbor," *nami* = "wave"). Tsunami is an obvious example, but there are many words that are more subtle examples of horizontal transfer. Horizontal transmission of words has had a particularly large impact on the English language. In the eighth century, Viking conquerors brought words from Old Norse such as "same," "skirt," and "skin" to the Old English speakers in the British Isles. The English language also was exposed to the "French" spoken in Normandy because William the Conqueror took over the British Isles in 1066. We have the Norman French to thank for such common words as "blue," "tax," "fry," "coast," and "park."

Note here that the slowly evolving vocabulary is the part least tied to changing culture, whereas the most changeable is the most culture-bound. Note that "park" and "tax" are related to government. The relation with Old Norse was a bit more complicated, as it is likely that for many decades in Yorkshire there was variation in pronunciation between a Norse "accent" with *sk-* and an English accent with *sh-*. Slowly, the Norse and English forms of the word took on slightly different meanings, as shirt and skirt, shatter and scatter.

The pervasive horizontal transmission of words across distantly related languages complicates determination of the relationships among those languages. Given the similarities between a large number of English and French words, you might think that English and French were closer to one another than each is to German if you focused on the wrong words. We know, however, that English is closer to German based both on the historical record and the similarities between English and German for a substantial portion of vocabulary. The lexicon of a language includes more than just its vocabulary. It also consists of each word's grammatical structure (what linguists call morphology), part of speech, and sound structure (phonology). Looking at the entire lexicon, English clearly is more similar to German than either is to French.

The methods biologists and linguists use to reconstruct common ancestors of species and languages are quite similar. These methods break down for the reconstruction of common ancestors of languages that are more than a few thousand years old, however. But these methods also break down for divergent groups of organisms when information from genes that evolve rapidly is used in the analysis. The fact that some genes evolve very slowly permits this type of analysis for even very divergent organisms (for example, plants, animals, and yeast). Moreover, in contrast to the opening quotation by Darwin, even if one were given a perfect pedigree of humanity, the patterns of genes might not exactly match the pattern of languages due to the horizontal transmission of words. Joseph Greenberg (1915–2001), the late Stanford linguist, had developed controversial tools to cope with the most distantly related languages. More than anyone else, Greenberg searched for a common ancestor for all of the world's 5,000 languages. We explore his contributions in the next section.

■ Joseph Greenberg and the Quest for Language Universals

Only Noam Chomsky ranks higher among twentieth-century linguists than Greenberg.[9] Born in Brooklyn in 1915, Greenberg was a prodigy in both music and linguistic ability. Before finishing high school, he had taught himself Hebrew and Greek. In addition to picking up Latin and Arabic in college, Greenberg studied Native American languages. If that weren't enough, Greenberg also spent his third year of college doing fieldwork in Nigeria, where he picked up the language Hausa.

Stationed mainly in Northern Africa during World War II, Greenberg engaged in breaking Italian codes as part of the Army Signal Corps. After the war, Greenberg began work on his first major accomplishment: classifying the 1,500 languages of Africa. It was Greenberg who placed the Khoisan languages into one of four superfamilies of languages; the other three were the Afroasiatic, the Niger-Kordofanian, and the Nilo-Saharan.

Greenberg then turned to classifying the languages of the Native Americans. He divided these languages into three superfamilies. Far north were the Eskimo-Aleut languages, of which Greenberg recognized just nine. Greenberg grouped 34 languages that are spoken mainly in what is now Western Canada into what he called the Na-Dene. The Apache and Navajo languages belong to the Na-Dene group; about 1,000 years ago, these tribes broke away from their more northern kin. The remaining 500-plus languages were placed into a single superfamily, the Amerind.

Greenberg's classification of the Native American languages was much more controversial than his prior classification of the African languages, and remains so to this day. In contrast to the three superfamilies of Greenberg, other linguists recognize dozens of families of Native American languages. Greenberg's critics charged that Greenberg's approach lacked rigor, and that several reasons other than common ancestry could account for words being similar in disparate languages.

Examining of the controversy requires exploring some of the details of Greenberg's methodologies. To achieve his aim of finding common ancestors of very distant languages, Greenberg developed a method called mass lexical comparison. This method differs from traditional comparative methods in several ways. The traditional method was to find specific "sound laws" that established equivalences between words in two languages (or protolanguages) by showing regular correspondence between sounds. Later in the 1940s, Morris Swadesh would compute similarities between the lexicons of two languages to determine relationships. In contrast, Greenberg argued for computing similarities between the lexicons of two sets of closely related languages. (This is the "mass" aspect of the mass lexical comparison.) To give an example, in Greenberg's method, one would look for relatedness of languages by comparing the lexicons of the Germanic languages (as a group), those of the Romance language (as a group), and those of the Cyrillic languages (as a group). Relatedness would be determined by which of these sets were most similar. Greenberg argued based on statistical principles that similarities calculated based on sets of lexicons would be more reliable than those calculated from just two languages.

Traditional methods of reconstructing language family trees (phylogenies) required intense study of each particular language. Such methods are exceedingly difficult and time-consuming when there are numerous languages to be studied. In contrast, Greenberg's method could reconstruct language phylogenies with just samples of the lexicons from each of the languages.

After being chair of the linguistics department at Columbia University, Greenberg moved to Stanford University in 1962. That move was serendipitous, as a young Italian geneticist named Luca Cavalli-Sforza would also move to Stanford. Cavalli-Sforza had shown the effects of genetic drift manifested by varying frequencies of genetic variants in different Italian villages. He also had used, and continues to use, numerical techniques and a large number of blood group genes to demonstrate varying influences on genes in Europe. For instance, Cavalli-Sforza and his colleagues found genetic signatures of the spread of agriculture across Europe.

Cavalli-Sforza showed that many of the language families that Greenberg had postulated—including those for the Native American languages—correspond with genetic patterns. For instance, genetic data support Greenberg's claim of only three distinct migrations colonizing the Americas.[10]

Greenberg, like Darwin, was interested in determining whether all languages were derived from one common language, with the eventual goal of reconstructing it. In support of such a common language of all, Greenberg noted the existence of some apparent similarities across nearly all languages. For instance, in nearly all languages, the root word "trik" is used for the words "one," "finger," "point," or "indicate." The exceptions happen to be the Khoisan and Niger-Kordofanian families.

■ Back to Click Languages

Joanna Mountain, a former graduate student of Cavalli-Sforza who returned to Stanford as a member of the faculty, has continued the work of her mentor of linking language and genetic evolution. With Alec Knight, a postdoctoral fellow in her lab, Mountain has investigated the evolution of the click languages in Africa. Mountain and Knight were particularly interested in the extent to which different populations of click language speakers are genetically differentiated. The geographical distribution of Khoisan languages suggested that some click speakers might be quite distinct from others. Although the speakers of these languages are mainly concentrated in the Kalahari Desert, some Khoisan speakers—the Hadzabe, for instance—live in Tanzania, several hundred miles to the north and east. To explore this possibility, Knight, Mountain and their colleagues sequenced mitochondrial DNA and Y chromosome DNA from several groups in central and southern Africa, including some click language speakers.

Knight, Mountain, and their colleagues found that the click speaking !Kung and the Hadzabe are genetically as distant or more than any other pair of African populations.[11] This finding was based on both mtDNA and Y chromosome data, and thus both the all-paternal and the all-maternal lineages support an ancient split between these two groups. The upper bound of the split between the Hadzabe and the !Kung is 110,000 years (plus or minus 40,000), but this estimate is based on problematic assumptions. Regardless of the exact date, we can say with confidence that the separation is tens of thousands of years old.

So how can we explain the existence of clicks in languages spoken by populations that split so long ago? It seems very unlikely that clicks were invented independently in the languages spoken by various peoples in the Kalahari and by the Hadzabe in Tanzania. First, clicks are very rare in general, and thus we would not expect their invention to be a frequent occurrence.

Another argument against both groups acquiring their assemblages of clicks independently is that repertoires are rather complex, and complex inventions are also not likely to be frequent events. Moreover, click repertoires of these geographically distinct groups overlap considerably.

Several other hypotheses remain that explain the similarity of clicks in these languages spoken by people with substantial genetic differentiation. Suppose that the Hadzabe population had actually diverged only fairly recently from the other Khosian-speakers, such that their languages were fairly similar and retained the clicks. Further suppose that after the split, the genes of the Hadzabe had been replaced by those of another population, but that the Hadzabe language remained mostly intact. Such a process of gene replacement without language replacement is possible and would account for the discrepancy between the linguistic and the genetic patterns that Knight and Mountain found. The problem with this scenario is that it is not very parsimonious (chapter 3); it rests upon several assumptions.

Another possibility is that the Hadzabe acquired clicks from another Khoisan-speaking population. Such linguistic borrowing isn't that far fetched; recall that the Bantu language Xhosa acquired clicks from one of the Khoisan languages.

The most intriguing possibility is that the Hadzabe and the !Kung really did diverge tens of thousands years ago, and that the clicks have remained all this time. This scenario raises the question: why have clicks been maintained in a small group of languages for so many thousands of years?

It's possible that clicks are neutral characters, neither advantageous nor harmful. We have seen that genetic drift can cause neutral variants of genes to be lost in some populations but become fixed in others. A similar process could have occurred with respect to language; clicks have been fixed in the Khoisan languages, but lost in virtually all other languages simply due to random processes.

More intriguing is the notion that clicks are adaptive in certain circumstances. When stalking prey, the !Kung speak almost exclusively in whispered clicks. Thus these clicks may have had adaptive value in hunting societies, and the selective pressure would have maintained them in the lexicon. In contrast, pastoral societies would have little use for the clicks, and over time, they would erode.

■ Recommended Reading

Here are several excellent recent books that provide general background on language evolution. The books by Cavalli-Sforza also discuss the relationship between genetic and linguistic evolution. Pennock's book, though mainly about the rise of Intelligent Design creationism, discusses language evolution in detail. Pennock draws several analogies between the two types of evolution.

Cavalli-Sforza, L. L. 2000. *Genes, Peoples, and Languages.* University of California Press.

Cavalli-Sforza, L. L., and F. Cavalli-Sforza. 1995. *Great Human Diasporas: The History of Diversity and Evolution.* Helix Books.

McWhorter, J. 2001. *The Power of Babel: A Natural History of Language.* W. H. Freeman.

Pennock, R. T. 1999. *Tower of Babel: The Evidence Against the New Creationism.* The MIT Press.

Who Let the Dogs in?
The Domestication of Animals and Plants

The first and chief point of interest in this chapter is, whether the numerous domesticated varieties of the dog descended from a single wild species, or from several.

—Charles Darwin (Darwin, 1868 [1998], p. 15.)

■ The History of Domestication

Until about 15,000 years ago, nearly all of human groups were nomadic hunter-gatherers. Environmental changes, including the retreat of glaciers and the diminishing supply of large mammals, led to some human groups becoming more sedentary. The persistence of human populations was a necessary but not sufficient step in the development of agriculture. But even prior to the practice of full-scale agriculture, people were beginning to domesticate animals and plants.

Agriculture first developed in the Fertile Crescent, a region that extends from the Sinai Peninsula past the Tigris-Euphrates River valley in present-day Iraq. Archaeological evidence from the Abu Hureyra site in Syria establishes that the earliest farmers probably started cultivating wheat and rye after a severe drought approximately 13,000 years ago. Although agriculture did spread within and out of the Fertile Crescent, it also was later independently established in several other places across the globe.

The reasons behind why humans adopted agriculture are fascinating in their own right, but they are not the primary focus of this chapter. Instead we explore questions related to what DNA markers and the application of population genetics principles can tell us about the animals and plants that were domesticated. When were particular animals and plants domesticated? What were the wild progenitors of the domesticated species, and how many origins were involved? Did the domesticated species undergo a genetic bottleneck during the process of domestication? What were the effective population sizes? Is each domesticated species one big randomly-mating population, or is there genetic structure in those species? These are the types of questions that can be addressed by Darwinian detectives applying molecular genetic markers and evolutionary theory.

Discussion of these questions for all or even just the most important domesticated species could well fill several books. Thus, this chapter will focus on a few aspects about the domestication of dogs and corn.

■ Just What Is a Dog?

Darwin, as shown in the quotation that opens this chapter, was not sure from how many species dogs had originated. Perhaps overly pessimistic, Darwin also stated, "We shall probably never be able to ascertain their origin with certainty."[1] The great diversity in size, shape, color, temperament, and other characteristics among dog breeds led Darwin, and many others since, to suspect that dogs had multiple origins. Could Newfoundlands and Chihuahuas, Pit Bulls and Cocker Spaniels, Whippets and Australian Shepherds, and all of the other breeds really come from a single species? Moreover, the fact that dogs can produce hybrids with a number of species—wolves, coyotes, and jackals—also suggests that dogs had multiple origins.

As we shall see, molecular studies strongly support a single species—the wolf—as the progenitor of the domestic dog, but they also suggest that dogs came from multiple populations of wolves that were widely separated in space (and probably time).

Before we consider in detail what the molecular data tells us about dog domestication, let's discuss two alternative views pertaining to how dogs were domesticated. The traditional view is that dogs are domesticated wolves; humans tamed wild wolves, then subjected these tamed wolves to artificial selection and in the process created the Poodle, the Collie, and all other breeds we know and love.

Ray Coppinger, who in addition to being a biologist at Hampshire College is also dog sled trainer, put forth a most intriguing alternative hypothesis about the evolution of dogs.[2] In contrast to the view that humans domesticated dogs, Coppinger argues that dogs domesticated themselves.

Coppinger's hypothesis begins on well-established grounds that starting around 15,000 years ago humans began living in permanent or semipermanent villages in greater numbers. He argues that such settlements created a new niche—the garbage dump! At these dumps, wolves would hang around and scavenge for food. Wolves tend to be nervous and skittish—qualities that would not benefit them as garbage-seekers. Those wolves, however, that were somewhat less skittish than others would be favored in the garbage dump; and to the extent that this trait had a genetic component, natural selection would favor less skittish wolves. As they became more tolerant of and less

Figure 11.1

Examples of modern domesticated dogs. *Top*: A Labrador retriever (yellow), named Cousteau. Courtesy of Sharlene Santana. *Bottom left*: A Boston terrier, named Jenkins. Courtesy of Kate Wells and Chad Hoefler. *Bottom right*: Lucy, described by her owner (Kara Belinsky), as "a complete mutt—mostly pit bull, and Australian cattle dog with some chow chow and German shepherd." Courtesy of Kara Belinsky.

nervous around humans, these wolves—now proto-dogs—were spending more time with humans. In the course of time and further natural selection, the resulting animals became like the village dogs that still haunt dumps across the world. These dogs were then in a position to be subject to artificial selection. Most of the variation in breeds that we see is obviously the result of artificial selection that we have imposed, but the bulk of this artificial selection occurred recently; the initial domestication did not involve artificial selection.

Definitive proof of Coppinger's hypothesis has not been found, nor has the traditional explanation been refuted. Nonetheless, a couple lines of evidence seem more compatible with the Coppinger's hypothesis than with the traditional explanation.

Coppinger presents as evidence for his view the dogs in Pemba, an isolated island off the East Coast of Africa. The Pemba people dwell figuratively at the edge of the Agricultural Revolution; although they subsist mainly as hunters and gathers, they also fish and raise cattle and chickens, as well as various vegetables. These people don't keep dogs as pets; in fact, they have a dislike of dogs based on religious taboos. Yet dogs run through Pemba villages, eating scraps, much as pigeons do in American parks.

Coppinger describes these dogs:

> They all looked just about the same—thirty or so pounds, slender, with short, smooth variegated colors, some with large spots, some with markings on their heads, ears, legs, or tails. Their ears are pendant, or erect but bent over slightly at the tips.[3]

It is dogs like these found in Pemba that Coppinger sees as the "missing link" between wolves and dogs. Such village dogs, Coppinger points out, can be found in "undeveloped" areas all over the world.

Though suggestive, the Pemba dogs do not prove Coppinger's hypothesis. We should keep in mind a couple caveats. First, the fact that Pembans don't keep dogs now doesn't mean they didn't in the past. Moreover, many peoples don't keep pets as we in the west do—and a dislike of dogs is hardly unique to Pemba.

Coppinger also points to a demonstration experiment examining correlated responses in morphology that arose in response to selection for behaviors associated with domestication in foxes that mirror the differences between dogs and wolves. Darwin, who knew nothing of genes but understood the essence of heredity, knew that characters do not exist in a vacuum and that selection on one character can often affect multiple characters. He called this the "law of correlation." Modern evolutionary quantitative geneticists, who

Figure 11.2
Village dog from
Senegal. Such village
dogs can be found all
over underdeveloped
regions of the world.
Courtesy of Ray
Coppinger.

speak of genetic and phenotypic correlations, have built theoretical frameworks to describe and predict such correlated responses.

This experiment came from an unlikely source: a Siberian scientist named Dimitry Belyaev, during the tail end of Lysenkoism. Recall that under Lysenko, Mendelian genetics had been essentially outlawed and a generation of Soviet geneticists had been purged in favor of political orthodoxy. Belyaev, at the Institute of Cytology and Genetics in Novosibirsk, would initiate an experiment (under the subterfuge of studying fox physiology) selecting for tame behaviors in foxes, a species that had never been completely domesticated. This experiment, which has continued many years since Belyaev's death, would eventually demonstrate dramatic morphological changes arising as correlated responses to selection for tameness.[4]

Belyaev and his associates collected foxes (100 vixens and 30 males) from a commercial farm in Estonia and performed a selection experiment on them. Although these foxes were somewhat tamer than the typical wild fox, they were certainly not tame. Only one criterion was used in the selection process: how tame the foxes were. The researchers assessed tameness by subjecting pups to a battery of tests to see how they responded to humans and to other pups. No other factors were considered in choosing which foxes would contribute to the next generation and which would not. Selection was very intense—usually only the most tame 5% of male offspring and 20% of female offspring would be selected in any generation.

Of special interest is the most-tame category (Class IE), described as "the domesticated elite." Foxes in this category "are eager to establish human contact"; they engage in "whimpering to attract attention..., sniffing and licking experimenters like dogs."[5] None of the initial population, nor any of the foxes from the first five generations of selection, fit into this category. Even after 10 generations of selection, only 18% of fox pups were classified as domesticated elite." By the late 1990s (after between 30 and 35 generations), however, 70 to 80% of the foxes were considered elite. This is yet another example of the power of selection to transform populations in a relatively short period of time.

The "domesticated" foxes in Belyaev's experiment changed in numerous ways. In addition to reaching sexual maturity about a month earlier than wild foxes, the "domesticated" foxes also produce litters that exceed those of wild foxes by one pup on average. Sexual dimorphism is reduced in the foxes, and the males seem to have been "feminized"; that is they look more like wild vixens. In both sexes, the heads are smaller, both in width and height, and they tend to have shorter and wider snouts. The changes in the head size as well as the feminization are consistent with changes that occurred in other species that have been domesticated, including dogs.

One of the most striking changes in the "domesticated" foxes is the presence of a piebald coat, in which regions of the coat lack pigmentation, resulting in pure white fur and pink skin underneath. Piebald coat color is often found in domesticated animals, including dogs, horses, pigs, and cows.

Floppy ears and shorter tails—as found in dogs—also appeared in some of the "domesticated" foxes.

So what can molecular data tell us about the when, where, and how of dog domestication? In addition, do these data support or contradict Coppinger's hypothesis?

■ Did Dogs Arise 100,000 Years Ago?

An early mitochondrial DNA study of dog origins shocked the scientific community because it proposed that dogs as a separate lineage were far older than anyone had previously thought. Robert Wayne's group at UCLA analyzed mitochondrial DNA data, and concluded that dogs could have originated as much as 135,000 years ago—a date ten times as old as the archeological evidence had suggested.[6] How did Wayne's group arrive at that date, and what else did their study find?

Wayne's group had obtained sequence from a rapidly evolving region of the mitochondrial DNA from a total of 140 dogs from 67 breeds, as well as 162 wolves from 27 different populations found in Europe, Asia, and North America. They found that dogs and wolves had strikingly similar levels of genetic diversity, both in terms of numbers of different haplotypes and the average distance between any pairs of haplotypes.

The data from Wayne's group showed that wolves are unambiguously the ancestors—and the only ancestors—of dogs. This inference was reached from data showing that the DNA sequence of dogs is far closer to that of wolves than it is to the sequence from any other wild animal. Any given dog was separated from any given wolf by a maximum of 12, and often many fewer, nucleotide changes. In contrast, dogs and coyotes were separated by more than 20 changes in the sequence; the same is true with dogs and jackals.

Wayne's group was unable to assign these mitochondrial DNA haplotypes to breeds; many breeds shared haplotypes, and several haplotypes were maintained within a few single breeds, including German Shepherds and Golden Retrievers. Haplotypes, however, did cluster into four distinct clades. The largest cluster (Clade 1), comprising 19 of the 26 haplotypes, included many common breeds as well as a few ancient breeds, such as the Basenji and the Greyhound.

Wayne's group obtained the estimate of the age of dogs by observing that the furthest distance between dogs within Clade 1 was 0.135 times the divergence of wolves and coyotes. Assuming that wolves and coyotes diverged one million years ago and applying the molecular clock, this yields 135,000 years. This estimate is an upper limit, and it rests on a number of assumptions. First, it assumes that the dogs in Clade 1 came from a single ancestral population that had little or no variation, and that all the existing variation within Clade 1 arose from new mutations and not from variation preexisting in the founding population. If the dogs in Clade 1 had actually come from several populations of wolves, this would make dogs appear older than they really are. Variation within the founding population would have a similar effect.

Wayne's estimate also assumes constancy in the rate of evolution among the different nucleotide sites. If substantial heterogeneity of evolutionary rates exists among the sites, then the clock assumption would be violated—the age of dogs would be younger than the estimate.

A third concern about this estimate is that it is based on a single gene that has the unique characteristic of being transmitted only through maternal lines. Remember that different genes can have different histories, and the genealogy determined by a single gene may not reflect the true history of the lineage. We'll soon see that autosomal genes give a different picture of the genealogy of dogs than does the mitochondrial DNA.

The results of a follow-up study do point to a much lower age for the divergence of wolf and dog, as well as point to southeast Asia as the most likely site for dogs' first origin. This study was led by Peter Savolainen, whose original interest in dog genetics arose as a way to assist crime-scene investigators in their analysis of hairs. His group found that four major groups of dog haplotypes and one isolated dog haplotype are interspersed with the wolf haplotypes.[7] These data allow us to infer that dogs are derived from at least five female wolf lines.

In each of the major clades, more haplotypes are found in East Asia than from any of the other regions. This is consistent with a hypothesized East Asian origin of dogs, because we expect more diversity to accumulate in the geographical region where dogs have been around the longest. By that same logic, we would expect the cluster with the most diversity to be the oldest. If we were to assume just a single origin of this cluster (Clade A), then its most likely age is 41,000 years. The problem is that the genealogical structure of this cluster suggests that it probably had multiple origins. Had it but one origin, it would have had a genealogical structure that resembles a starfish, with unique and rare haplotypes branching off from one center haplotype that is most frequent. In actuality, Clade A lacks a well-defined center haplotype. With the assumption of multiple origins, Clade A probably originated about 15,000 years ago. This latter date is in accordance with the archaeological data.

■ Beyond Mitochondrial DNA

A team at Seattle's Fred Hutchinson Cancer Research Center headed by Elaine Ostrander has been examining the genetic structure of dog breeds, in part as a means to find DNA variations in dogs that are associated with diseases. This group has been particularly interested in malignant histiocytosis, a form of cancer of white blood cells. Although generally very rare in dogs, this cancer is much more common in Bernese Mountain Dogs and Flat Coated Retrievers than in other breeds.

Heidi Parker, then Ostrander's graduate student, initiated a study using microsatellite DNA markers to investigate genetic variation in dogs.[8] Microsatellites, which are very short repetitive sequences of DNA, have often been used to examine genetic variation of a variety of organisms. One major advantage of microsatellites is that they quickly provide information from many different regions of the genome.

Parker, Ostrander, and their colleagues surveyed 414 purebred dogs representing 85 breeds at 96 microsatellites. They found that purebreds, if treated as a single population, have an overall genetic diversity that is comparable to that of humans. Unlike humans, however, dogs are highly structured genetically. That is, the vast majority of the variation in humans is among individuals within populations; in contrast, 30% of the genetic variation in dogs based on the microsatellite data is among-breed differences, far higher than so-called "racial" differences among humans.

Remarkably, enough structure is present among purebred breeds such that a computer program, into which just the DNA profile was fed, could assign dogs to breeds with remarkable accuracy. Only 4 of the 414 dogs were incorrectly assigned to breeds based on their microsatellite profile.

A phylogenetic tree of the microsatellite data indicates that the deepest split included four African spitz-type breeds: the Shar-Pei, the Shiba Inu, the Akita, and the Chow Chow. The inference to be drawn from this tree is that these breeds are the oldest of existing breeds. This result suggests that dogs originated in Africa, in contrast to the results from the mitochondrial DNA study. Because the microsatellites reflect the genome as a whole, and not just a single gene, the inferences drawn from the microsatellites are likely to be better supported than those based on the data from the mitochondrial DNA. Thus, dogs probably had an African origin.

■ The Dog Genome

According to the Chinese calendar, the most-recent year of the dog began in late January 2006. The previous month, the complete sequence of the dog genome was published in *Nature*.[9] Chosen for the project was Tasha, a female Boxer. Samples of the genome from a variety of other dog breeds were also sequenced to obtain information about patterns of genetic variation and linkage disequilibrium (LD), the correlations of genetic variants, in dogs.

At 2.4 billion nucleotides, the overall genome size of dogs is somewhat lower than that of humans and just slightly smaller than the mouse genome. The differences in genome size are probably due to loss of genetic material in the dog lineage. Along the dog lineage, genes have evolved slightly faster than in the human lineage, but much slower than the rate in the mouse lineage. One likely reason for these differences is the longer generation times—human generations are longer than those of dogs, which in turn are far longer than those of mice. Another possible explanation is that mice have faster metabolic rates than do dogs, which have faster metabolisms than do humans. Still another possibility is that parts of their genome were lost as dogs were domesticated.

Inferences about the evolutionary history of dogs can be obtained from examinations of the correlations of genetic variants—linkage disequilibrium (LD)—seen in dogs. LD is a measure of the extent to which knowledge of which genetic variant a particular dog has at a particular location provides predictive power about what genetic variants that dog will have at a different site in the genome. Recall that in humans, LD values for sites that are within a few thousand nucleotides are quite high, meaning that knowing what variant an individual has at one site is a good predictor of what variants that individual will have at other nearby locations. Beyond a few tens of thousands of nucleotides, LD usually rapidly declines in humans when the correlations among more distantly located genetic variants approach zero. The extent to how far apart on the chromosome LD persists is influenced by the extent of the effective population size of the species. LD should remain high longer in species with high effective population sizes than in ones that have had their effective population sizes reduced by genetic bottlenecks.

When dogs of all breeds are examined as a whole, the patterns of their LD are similar to those of humans. In fact, the LD of dogs drops off somewhat faster with genetic distance than does the LD of humans. This suggests that as a species as a whole, humans have experienced more intense genetic bottlenecking than have dogs. When individual breeds are examined, however, a different pattern emerges: LD drops sharply for the first 100,000 nucleotides or so, but then remains high for several million nucleotides.

The patterns of LD support the scenario of two major bottlenecks in the genetic history of most dogs, one upon the domestication of dogs and a subsequent one upon the strong selection that occurred during the creation of most breeds. The short-range LD is the signature of the early bottleneck, and the longer-range LD reflects the bottlenecks associated with the creation of the specific breeds. Different breeds show different extents of long-range LD, with, for instance, Akitas and Rottweilers having high long-range LD and Irish Wolfhounds having lower long-range LD. These differences probably reflect differences in the extent of the bottleneck in the creation of the breeds. Labrador retrievers, which have far less long-range LD than the other breeds of dogs, have maintained a large population size and are probably the result of considerable mixing.

A decade of molecular evolutionary research on dogs and their relatives has already provided us with a rich understanding of their origin and evolution. The molecular data are consistent with the picture of wolves evolving around the time humans started living in semipermanent villages and in several different localities. The signatures of this bottleneck and the bottlenecks associated with the formation of separate breeds can be found in the patterns of DNA sequences. Further studies and advances will almost certainly reveal much about "man's best friend" over the next few years.

Dogs are far from the only domesticated species for which molecular evolutionary studies provide information about the natural history of their wild progenitors and the unnatural history they have had under domestication. We turn next to maize, a plant that has changed from its wild ancestor almost as dramatically as breeds of dogs differ from wolves and one another.

■ How Was Corn Domesticated?

Maize (*Zea mays*, colloquially referred to as corn; I will use the terms maize and corn interchangeably) is perhaps the most striking case of genetic modification associated with domestication. Although one can clearly see the resemblance of dogs to wolves, and other domesticated animals and plants as being related to their wild progenitors, it is much harder to discern that corn came from teosinte, a wild grass native to the Americas.

The question of maize's origins has interested biologists for many decades. Nikolai Vavilov, the Russian geneticist, who died in prison in 1943 after being repressed by Lysenko, was among the first to propose a hypothesis for maize origins. He argued that maize probably arose near Guadalajara, Mexico because, in this species, more variation is located in that region than elsewhere in the world.

Perhaps the most striking difference between corn and its wild relatives is in the floral architecture. Each individual plant of corn has but a single stalk and ear, but wild teosinte plants have multiple stalks and ears. The change was probably to concentrate effort into one single product.

Other important changes between teosinte and corn are also evident. For instance, think of the part of corn from corn on the cob that sticks to your teeth. This part, which surrounds the kernel, is called the glume. Teosinte produces a much larger and much harder glume than does domesticated corn. The glume is an adaptation in teosinte; it evolved to protect seeds when they passed through the digestive tracts of mammals. Such a hard glume,

Figure 11.3
Domesticated maize and teosinte. Courtesy of John Doebeley.

however, is not advantageous in a seed crop because it makes it difficult for humans to chew and digest the kernel. Thus, in the process of domestication, artificial selection was imposed to reduce the glume.

Maize comes from teosinte, but what is teosinte? In fact, teosinte is not a single entity, but rather several species of these grasses, which are all in the genus *Zea*. Included in the teosintes are two widespread annuals; in addition to *Zea mays*, which will we come back to in a minute, there is also *Zea luxurians*, found in Central America. Two other teosintes are perennial species isolated to what is now Jalisco, Mexico: *Zea diploperennis* and *Zea perennis*. The latter species is interesting in that its cells contain four sets of chromosomes, instead of just the typical two.

Complicating the situation further, four recognized subspecies exist within the species *Zea mays*. In addition to domesticated maize *(Zea mays mays)*, two Mexican subspecies are known: *Zea mays mexicana*, which lives at higher elevations and has large spikeletes (the small flowers that make up the male inflorescence), and *Zea mays parviglumis*, which lives at lower elevations and has small spikeletes.

Domesticated maize also varies considerably in morphology, as well as in its number and form of chromosomes. This variation has led many to speculate that corn may have had multiple origins. But as we shall see, DNA evidence rules out the hypothesis of multiple origins for maize.

John Doebley at the University of Wisconsin has been applying molecular markers and the principles of population genetics to studies of maize and its wild relatives.[10] To address the question of the origin of maize, Doebley and his colleagues collected 193 maize plants from all over its pre-Columbus range (southeastern Canada to Chile). They also collected a total of 67 plants from both *mexicana* and *parviglumis*, plus smaller numbers of the other teosinte plants to be used as outgroups. Their next step was to ascertain the genotypes for all of these plants at a total of 99 microsatellite loci more or less randomly placed across the genome.

The results from this study strongly support corn's being derived from *parviglumis* and *parviglumis* alone. This conclusion rests upon the fact that all haplotypes from domesticated maize plants cluster together within the *parviglumis* clade. Not only does this result support a *parviglumis* origin for corn, but it also suggests that the domestication originated in a single population of *parviglumis*.

The phylogeny also provides clues to the probable first place where maize was domesticated. The earliest lineages of domesticated maize are from the Mexican highlands, so it appears that first domestication occurred there. This technology then spread to the lower elevations, and then throughout the eastern part of the Americas.

You may have noticed a possible inconsistency; although maize was probably first domesticated in the Mexican highlands, *parviglumis* (its most likely ancestor) is a lowland species. How can this be? One possible explanation is that *parviglumis* once had a larger range, which became restricted with time.

It is also possible that the first people to domesticate maize actually moved *parviglumis* to the highlands when they domesticated it.

Doebley's group was also able to estimate the date of the domestication. Some of these microsatellite markers with known mutation rates evolve in a fashion that makes them suitable to be used for dating. Analysis of the data from these markers reveals that the divergence between Mexican maize and teosinte probably occurred 9,200 years ago, but could have been as recently as 5,700 years ago or as long ago as 13,100 years. Although the oldest known fossil maize dates to 6,250 years ago, some archaeological evidence points to maize domestication beginning considerably earlier.

This study and a subsequent one also provided evidence of a current or very recent gene flow among *mexicana, parviglumis*, and maize. This gene flow, although detectable by using population structure analyses, is at a low level. Of particular interest, genes from *mexicana* are getting into the maize of the Mexican highlands.

■ Morphology, Genes, and Domestication

Starting in the 1990s, biologists interested in the evolution of morphology made substantial progress in determining the actual genes involved in the morphological change, as well as their evolutionary trajectories. For these biologists, domesticated species offer advantages because within them a considerable amount of change has occurred in a relatively short time. The list of specific genes that influence morphology in a number of different domesticated species, including maize, continues to grow.

Teosinte Branched1 (Tb1) is a gene that affects the morphology of maize. Mutations that knock out its function cause maize to look more like teosinte. In particular, these mutant plants have longer lateral branches with tassels at the ends, instead of the typical short branches that end in ears. At the biochemical level, the Tb1 protein acts to repress growth of the organs wherein it is expressed. Plants that carry the maize variant of *Tb1* express it heavily in the precursors to the lateral branches, thus stunting their growth. In contrast, plants with the teosinte variant express less of the Tb1 gene in their branches, allowing them to grow longer.

John Doebley's group found several lines of evidence for a selective sweep occurring in the regulatory region of the *Tb1* gene, one of which is the pattern of genetic variation observed.[11] In this region, little genetic polymorphism exists within maize plants, but teosinte plants display much higher levels of variation, suggesting a selective sweep in maize. Moreover, the part of the gene that codes for the protein shows normal levels of variation in both maize and teosinte, thus suggesting that the selection involved concerned not the structure of the protein but the regulation of its expression. The data collected by Doebley's group also enabled them to estimate that the selection at the Tb1 gene's regulatory region took place during a period that lasted between 300 and 1,000 years.

■ Agriculture's Effect on Human Social Structure

The introduction of agriculture was not an unmixed blessing. For instance, Jared Diamond and others have argued that one consequence of agriculture has been the proliferation of disease epidemics that had been virtually unknown in hunter-gatherer societies. The advent of agriculture has also probably affected human social structure. Can we see the impact of the shift to agriculture in patterns of human genetic variation? Apparently so. An Italian group led by Giovanni Destro-Bisol finds genetic evidence that hunter-gatherer and agricultural societies in Africa exhibit different sex-linked patterns of migration.[12]

Destro-Bisol's group examined Y-linked and mitochondrial DNA markers in both African hunter-gatherer populations such as the !Kung, whom we discussed in the previous chapter, and in agricultural populations in sub-Sahara Africa. Among the !Kung and the other hunter-gatherer societies, far less geographic clustering of variation is observed in the Y-linked markers than is observed from the mitochondrial DNA markers. The extent of geographic clustering is negatively correlated with the expected migration rates. The inference to be drawn from this result is that the estimated migration rate of males (as determined by the patterns seen in the Y chromosome) was much greater than the estimated migration rate of females (determined by patterns seen in the mitochondrial DNA). In fact, males migrated an estimated nine times more than females in these groups. The farmer societies, by contrast, exhibited slightly more geographic clustering in the Y-linked markers than the mitochondrial DNA markers; the estimated female migration rate was 1.5 times greater than that of the males. Thus, there is a 16-fold difference in the ratios of male to female migration rates between hunter gatherer and agricultural societies. It appears that at least in African populations, the advent of agriculture altered the patterns of migration from male-biased to female-biased. One plausible explanation for this change is that the inheritance of territory from father to son reduced male migration in the agricultural societies.

■ Recommended Reading

General information about domestication can be found below.

Diamond, J. 1998. *Guns, Germs, and Steel.* W. W. Norton.
Diamond, J. 2002. Evolution, consequences and future of plant and animal domestication. *Nature* 418: 700–707.
Smith, B. D. 1998. *The Emergence of Agriculture* (paperback edition). Scientific American Library.

12 ∎

Size Matters
Toward Understanding the Natural
History of Genomes

Although full-genome sequencing has revealed numerous patterns of variation in genomic architecture among major taxonomic groups, a formidable, remaining challenge is to transform the descriptive field of comparative genomics into a more mechanistic theory of evolutionary genomics.

—Michael Lynch (Lynch, 2006, p. 450)

Deciphering how genomes come to acquire their characteristics, and how these in turn affect the evolution of features at higher levels of organization, will demand a broadly integrative approach that synergies insights from various disciplines.

—T. Ryan Gregory (Gregory, 2005, p. 706)

∎ Variation in Genomes—A Survey

Throughout the course of this book, we have focused on genes and their natural histories. We have explored what can be inferred from patterns of variation in DNA sequences about the evolutionary forces that have acted on those genes and the organisms that harbor them. Here, in the final chapter, we turn our attention away from particular nucleotide sequences of genes to the properties and natural histories of genomes as whole entities.

Perhaps the most puzzling feature of genomes is the sheer variation in their size across the different lineages of life. The largest known genome is more than 100,000 times larger than the smallest known genome of any free-living organism. Some general trends do emerge: bacteria, which lack true membrane-bound nuclei and hence are considered prokaryotes, generally possess small, streamlined genomes. By contrast, organisms with true nuclei enclosed by membranes—the eukaryotes—have larger cells and larger genomes. Genomes of multicellular eukaryotes tend to be larger than those of single-celled eukaryotes. Although there are many exceptions, genomes are larger in vertebrates than in insects and other invertebrate animals. But, as we shall see, the relationship between genome size and complexity of an organism is at best weak and muddy.

Let's start with our own genome. The human genome contains just over 3 billion nucleotides. This figure is actually the size of the genome found in the gametes (sperm and eggs), and is known as the haploid number. In contrast, the cells in the rest of the body contain twice as much genetic material—one haploid set from both the mother and father; this is the diploid number. Except when mentioned otherwise, the genome sizes discussed here will be of the haploid number. To put this and other large numbers in perspective, let's consider genomes as genetic libraries filled with books that each consist of one million nucleotides. For the sake of comparison, the book you are reading has about 600,000 characters. By this analogy, the human genome would contain slightly over 3,000 books filled with a million characters each—comparable to a very large, but not outrageously so, personal library. A person reading a book a week from age 10 to age 70 would read the equivalent of the human genome.

Not much variation exists within mammals for genome size, at least not the same extent that we will see in some other groups; for instance, the mouse and rat genomes both are about 80% the size of that of humans. Nor is the range of genome sizes in birds large, at least not when it is compared with the extraordinary variation seen in the genomes of fish, or especially amphibians.

The smooth pufferfish *(Fugu rubripes)* is a popular Japanese delicacy, but if prepared incorrectly it can cause lethal poisoning because its liver and ovaries contain the neurotoxin tetrodotoxin. The genome of this pufferfish is also remarkable. At 370 million nucleotides—that is, only 370 books—*Fugu rubripes* and its closest relatives have the smallest vertebrate genome. The prolific science fiction and science writer Isaac Asimov published more than 370 books in his lifetime.[1] Granted that most of the books Asimov wrote each contained fewer than a million characters, but if one includes all of the short stories and other material Asimov also published, his collective works would almost certainly include more characters than the pufferfish genome. We'll discuss in more detail the extremely small size of the pufferfish genome, but there is no obvious reason that the pufferfish genome should be seven times smaller than the mouse genome.

On the other end of the scale, genomes of some fish and some amphibians exceed 100 billion nucleotides. One example of a fish with a gigantic genome is the lungfish. Closely related to coelacanths, lungfish are remarkable because they can breathe air through a modified air bladder in addition to through their gills. Limb-like appendages replace fins in these fish. As we will see later in this chapter, genomes of some salamanders approach this size. These genomes contain the equivalent of more than 100,000 volumes, comparable to the sizes of many libraries in small cities. Again, there is no obvious reason that the genomes of these salamanders and lungfish should be 30 times the size of the human genome and nearly 300 times the size of the smooth pufferfish genome.

The variation in genome size across vertebrates, for the most part, is not due to variation in the number of genes that characterize these animals. Only 1.5% of the DNA (45 million nucleotides—or only 45 books!) in the human genome codes for proteins. Within vertebrates, the amount of coding DNA remains relatively constant; somewhere between 30 and 60 books are reserved for coding DNA in almost all species of vertebrates. The same types and numbers of genes are present in vertebrates, including smooth pufferfish, humans, and salamanders. Instead, the variation in genome size relates to the extent of noncoding DNA.

Most insects have genomes that are smaller than most vertebrates. In most orders of insects, the genome size is between 150 million and 1.4 billion nucleotides, or between 5% and 50% of the human genome. Yet some insects contain comparatively large genomes. Genomes of many grasshopper species, for instance, can be as large as 15 billion nucleotides (15,000 genome books), five times larger than human genomes. Other exceptionally large insect genomes are found in stick insects (up to 8 billion nucleotides, or 8,000 books). The typical size of genomes in dragonflies is between 1.5 and 2 billion nucleotides, intermediate between the grasshoppers and most other insects in genome size.

Within flowering plants, genome size varies by at least a thousand-fold. The genome of the mustard plant *Arabidopsis thaliana* is only 125 million nucleotides (125 books), somewhat smaller than the *Drosophila* genome. This small genome has been fortuitous because it has made completion of the genome project for *Arabidopsis thaliana*, a plant so extensively studied by geneticists that it has been called the "green *Drosophila*." On the other end of the scale, genome size can be as large as 120 billion nucleotides (120,000 books) in lilies in the genus *Fritillaria*. Certainly lilies are not 1,000 times as complex as mustard plants; in fact, one would be hard pressed to say which of these plants was the more complex. As a general rule, genomes of annual plants are smaller than in the longer-lived perennials.[2]

Genome size is almost always essentially constant within species in both animals and plants, but there are exceptions. One exception is domesticated corn (*Zea mays*), wherein genome size varies by 40% among different individuals.[3] This variation can be seen in the chromosomes. Some individual corn plants carry extra small chromosomes, called B chromosomes, and/or extra pieces of DNA at the ends of chromosomes (called knobs). Perhaps this intraspecific variation is a remnant of corn's recent increase in genome size. The genome of domesticated corn is much larger than that of its wild relatives, and some biologists speculate that the larger genome (and an associated larger cell size) helped corn better respond to the colder climates and thus facilitated its spread northward during domestication. The increase in genome size as corn was domesticated is counter to the typical pattern; in most cases of domestication, genome size shrinks.

Outside of animals and plants, we find still more variation. Although genome size is usually smaller in single-celled organisms than in multicellular ones, many exceptions exist. Indeed, genomes of some single-celled amoebas are 200 times larger than the human genome!

■ The Selfish DNA Paradigm

Why do genomes vary, and do so by such large degrees? What is responsible for the distribution of genome sizes in different organisms? Do specific properties of organisms influence the size of their genomes, and vice versa? How does genome size relate to Darwinian fitness?

Many biologists during the 1960s and 1970s were keenly interested in these questions. The immense variation in genome size and the lack of correspondence between genome size and the complexity of organisms is a puzzle—one that we'll refer to as the genome size enigma. (Biologists often refer to the genome size enigma as the "C-value paradox," because C-value was a term coined by Hewson Swift in 1950 for the amount of DNA in the cell of a given organism; the C stood for constant.)

By the early 1980s, the genome size enigma appeared solved. The apparent solution was that most of genomes consisted of DNA of no purpose to the organism. Several discoveries led to this conclusion. Biochemical studies during the 1960s and 1970s showed that a large fraction of DNA in most organisms was repetitive, and sometimes highly repetitive. In the late 1970s, molecular biologists found that genes were broken up in pieces; the parts expressed in proteins (exons) were interspersed with parts that did not code for proteins (introns). Around the same time they found introns, geneticists also discovered pseudogenes, genes that were once functional but are no longer so. As we saw in chapter 3, these pseudogenes evolve quickly because their sequence is not constrained by negative natural selection. The same appears true of introns. In his 1976 book, *The Selfish Gene*, Richard Dawkins popularized the notion that genes existed not for the good of the species but because they helped their own persistence by building a well-functioning organism. In 1980, several biologists, including Francis Crick, the co-discoverer of the structure of DNA, published influential reviews that took Dawkins one step further; pieces of selfish DNA could promote their own replication with no benefit, and perhaps even a cost, to the organism. These selfish genetic elements would thus be parasites of the genome.

The selfish DNA paradigm views genome size as the result of competing forces. As Leslie Orgel and Francis Crick put it, "the amount of useless DNA in the genome is a consequence of a dynamic balance" between its accumulation and efforts by the cell to get rid of DNA.[4] According to the paradigm, organisms with genomes that have lots of excess DNA would be better off in terms of reproductive fitness if they got rid of it.

The selfish DNA paradigm for genome size is the prevailing but not universal view among contemporary biologists. To say that some of what we think of as junk or selfish DNA actually has a specific function is not a challenge to the selfish DNA paradigm for genome size, so long as most of the genome is considered selfish or junk. The key feature of the selfish DNA paradigm is that the optimal size of large genomes is much smaller than their current size, and that these genomes are large because they can't get rid of their junk DNA or prevent the spread of selfish DNA. Even those biologists who do not fully accept the selfish DNA paradigm acknowledge that much DNA is selfish or junk; the debate is over the extent to which larger genomes have been favored by natural selection.

Even if true, the selfish DNA paradigm does not really explain the genome size enigma. Saying that most of the genome is from genomic parasites or junk DNA does not explain why some genomes have more DNA than do others. What are the factors that determine how much excess DNA is found in genomes? Why do some genomes accumulate more junk, or are more tolerant of parasitic DNA, than others? These are the questions that need to be addressed to fully understand the genome size enigma.

We start by exploring one type of genomic parasite—transposable elements, otherwise known as "jumping genes."

■ Jumping Genes and Genome Size

The genome is not static; indeed, elements within the genome can copy themselves and change their positions. Such "jumping genes" are more formally known as transposable elements (and sometimes transposons or mobile DNA). They are ubiquitous and abundant; in fact, the total sequence length of these transposable elements accounts for almost half of the human genome. Let's briefly look at how these elements "jump" and how they behave in both large genomes (as in humans) and in smaller genomes (such as *Drosophila* flies).

Barbara McClintock, who received the Nobel Prize in 1983, discovered transposable elements in the late 1940s and early 1950s.[5] Having produced the first genetic map for corn and having been the first to show the relationship between recombination of genetic markers and the actual physical exchanges of genetic material on the chromosomes, McClintock was arguably the most important geneticist to work with corn. Observing genetic markers changing their positions led McClintock to the notion that genetic elements actively moved within the genome. Mobility within the genome ran counter to the tenets of 1950s genetics in which genes were like beads on a string. According to the then prevailing view, genes (as beads) could be exchanged between copies of chromosomes (as string) during recombination. Their positions on chromosomes could be changed via chromosomal inversions, but genes were viewed as stable; they did not jump. For two decades, McClintock's views were virtually ignored; but by the early 1970s, geneticists started observing mobile elements in bacteria and yeast. Not long after, the "jumping genes" that McClintock discovered in corn decades prior had been characterized and were shown to be similar to these mobile elements in single-celled organisms. By the time McClintock had received her Nobel Prize in 1983, transposable elements had been found virtually everywhere, including in humans.

The various transposable elements can be grouped into two major divisions.[6] In one, the elements produce an enzyme called transposase that, like the "cut and paste" function in a word processing program, cuts the transposable element out of its original position and pastes it elsewhere in the genome. How could this "cut and paste" action lead to an increased number of copies of the element? Several mechanisms are possible. One way in which elements can increase their copy number is if the jump occurs after DNA has been replicated but before the cell splits in two. So suppose that the element jumped in one of the chromosomes that had just been replicated, but not on the other. On the chromosome where the element jumped, a gap is left open from where the transposable element used to be. But there is no such gap on the copy of the chromosome in which the element did not jump. Within cells, repair mechanisms exist that restore damage done to DNA, and they often use the chromosome that lacked the gap as a template to fill in the gap on the chromosome where the element jumped. The consequence would be that on the chromosome where the element jumped, two copies of the element would appear—one in the original position and one in the new position.

The other major group of transposable elements uses RNA to increase in copy number. The transposable element DNA is transcribed into RNA; the information in that RNA, in addition to being translated into proteins, is also converted back into DNA information. The technical term for this conversion of RNA into DNA is reverse transcription—it literally is the reverse of transcription—and is mediated by an enzyme called reverse transcriptase. The newly reverse-transcribed DNA can be inserted back into the genome in different places. Such transposable elements, which are sometimes called retrotransposons, behave similarly to retroviruses, which also make use of reverse transcriptase. The most famous retrovirus is HIV, the virus that causes AIDS.

The jumping and copying—technically known as transposition—of both types of transposable elements is usually not frequent. Many elements jump as little as once every 100 generations or even once every 1,000 generations. Nevertheless, just as a dollar compounded at an annual rate of 7% interest will be worth over $20 million after 250 years, such infrequent jumping can have dramatic consequences over the long haul. Consider an element currently present on average in 20 copies throughout the genome per cell. Suppose that on average every element of this type jumps and copies only once per every 1,000 generations. Assuming that these elements are never eliminated and that they have no effect on fitness, the average cell will contain 200 copies of this element in its genome after 2,300 generations. If the organism is a fly, that is only a few hundred years! Unchecked, this element will reach an average copy number of 2,000 after another 2,300 generations. As Darwin noted, slowly reproducing elephants have the reproductive potential to take over the Earth. So do slowly jumping transposable elements.

There are 4 million copies of various transposable elements of various types lurking in the human genome. Together these elements constitute almost 50% of the total human genome. If all of the transposable elements in the human genome were laid end to end their total length would be equivalent to about 1,500 volumes! Two families of transposable elements are extraordinarily common in humans. The one million copies of the *Alu* element make up 11% of the human genome, and the half a million copies of the *LINE1* element make up 17%.[7] (The average length of individual copies of *LINE1* is larger than that of *Alu*.) In humans, most transposable elements are inactive; that is, they have lost the ability to transpose across the genome and are thus akin to fossils. Others are able to move but lack the ability to produce transposase or reverse transcriptase; they can move because other active elements produce the necessary enzyme. Roughly half of the elements in humans have appeared since the divergence from mice 80 million years ago. On average, there has been the addition of new 1,000 nucleotides every century. This rate of accumulation, however, appears to be slowing.

Transposable elements behave very differently in the fly *Drosophila melanogaster* than they do in humans (and in mammals in general). The activity of all transposable elements combined causes half of the visible mutations in this species of fly. We do not have a precise estimate for the comparable figure for the total effect of mutations from transposable elements in humans, but we do know that it is certainly far smaller than that in *Drosophila*. We do know that *Alu* elements—the most numerous group in humans—are responsible for only 0.3% of the mutations that cause disease in humans. One third of the mutations from *Alu* elements comes from the elements inserting themselves into genes leading to improper gene function, and the reminder arises from the improper recombination between *Alu* elements, resulting in duplications and deficiencies of genetic material.

In contrast with humans, in which most of the transposable elements are from only two families, *Drosophila* is host to many families of transposable elements. Geneticists working in *Drosophila*, keeping with the whimsical nature of *Drosophila* genetic nomenclature, have often named transposable elements with monikers suggesting the elements' peripatetic nature, such as hobo, gypsy, and jockey. The ship on which Darwin took a five-year voyage, the *H.M.S. Beagle*, is also the name of a transposable element found in *Drosophila melanogaster*.

One group of transposable elements, the *P* element, is especially active in *Drosophila melanogaster*.[8] This *P* element came to *Drosophila melanogaster* very recently via horizontal transfer from a distantly related species of *Drosophila, Drosophila willistoni*. This element was not found in strains of *Drosophila melanogaster* collected from the wild prior to 1950. *P* elements rapidly spread throughout populations during the 1960s and 1970s. In crosses where the male parent has *P* elements and the female lacks the *P* element, the *P* elements jump at much higher rates than they do in females with *P* elements. The result of the unconstrained jumping of *P* elements is high mutation rates, chromosomal rearrangements, and sometimes sterility in the offspring; a syndrome known as "hybrid dysgenesis."

As would be expected, *P* elements by themselves considerably affect fitness in *Drosophila melanogaster*. Studies suggest that the fitness of a fly with no *P* elements is 20% higher than the fitness of a fly burdened with the average number of *P* elements.[9] And yet because of their ability to jump and replicate in genomes and the phenomenon of hybrid dysgenesis, *P* elements spread in this species from very low frequency to high frequency in just decades.

A general pattern about transposable elements is emerging. In large genomes (such as found in humans), transposable elements are relatively stable, most copies belong to only a few families, and the fitness consequence of any one element is very small. In species with small genomes (for example, *Drosophila*), a much different pattern exists. Here transposable elements are usually active, come from numerous families, and often have measurable effects on fitness.

■ Kicking out DNA

Recent studies show that organisms possess mechanisms for eliminating DNA. The precise details of how these mechanisms work are not fully understood, and they are beyond the scope of this book. What is of interest for the discussion of genome size is that not only are organisms capable of getting rid of DNA, but also this capacity varies in different organisms.

Dmitri Petrov at Stanford University and his colleagues found that *Drosophila* loses DNA from defective transposable elements rapidly, at least compared with mammals.[10] Deletions in these elements occur 2.5 times as often in *Drosophila* as they do in mammals. These deletions also are about seven times longer in *Drosophila*, and thus these flies are kicking out about 20 times more DNA from defective transposable elements than are mammals.

The differential rate of DNA elimination in *Drosophila* and mammals is not due to the differences between mammals and insects; not all insects eliminate DNA at the same rate, and some get rid of DNA at a rate as slow or slower than that of mammals. Crickets of the genus *Laupala* live in Hawaii and have been the subject of work in the evolution of mating calls among and within species. This cricket's genome size, 1.9 billion nucleotides, is about 60% of the human genome and more than 10 times the size of the *Drosophila melanogaster* genome. Petrov and his colleagues find that DNA is lost in this cricket 40 times more rapidly than it is in *Drosophila melanogaster*.[11] Those species with more compact genomes appear to be better able to get rid of their junk DNA, whereas the species with larger genomes appear to be the equivalent of packrats.

There is a difference between the mechanistic reason and the evolutionary (or selective) forces for why genomes are particular sizes. That DNA excision occurs at a slower rate in *Drosophila* than it does in *Laupala* crickets is a mechanistic reason for why *Laupala* has a much larger genome than *Drosophila*. This mechanistic reason isn't the same as the evolutionary reason. It doesn't explain why there would be selection for maintaining a mechanism that purges excess DNA.

Petrov and his colleagues argue—quite convincingly—that each of these small deletions is essentially selectively neutral.[12] The amounts of DNA that are eliminated by any particular deletion are extraordinarily small in comparison to the total DNA in the genome, even in a small-genome organism like *Drosophila*. Thus, any effect that these deletions may have on fitness is almost certainly too small to be detected by natural selection. Petrov also presented data on the distribution patterns of these deletions that are consistent with individual deletions being selectively neutral.

Natural selection for smaller genomes could operate within a species via individuals possessing larger than average genomes tending to have a lower viability or reproductive success than individuals with smaller than average genomes. It is also possible that selection on genome size could operate at higher levels; if lineages with smaller genomes have lower probabilities of going extinct or higher rates of speciation, genome size would be checked. We will explore the latter possibility at the end of this chapter, for now we turn to exploring the effects of genome size on attributes of the organism.

■ How and Why Does Size Matter?

Many attributes of organisms are correlated either positively or negatively with genome size. But correlation is not causation. If we were to look at a group of children of different ages, we would find that height and vocabulary are highly correlated. Certainly, height is not the cause of vocabulary. Children's vocabularies don't increase because they are getting taller. It's even more ludicrous to claim that increasing vocabulary causes children to grow taller. Instead, increases in vocabulary and increases in height are due to separate processes of development. If, in the previous example, the children are all approximately the same age, the correlation between height and vocabulary size should largely (if not completely) disappear. It is possible that even with children of the same age, some correlation would persist due to the children being exposed to different environmental conditions affecting height and learning vocabulary in the same direction. But, even in this hypothetical case, height is not a causal factor for vocabulary increase (or vice versa).

Genome size strongly correlates with both the size of the nucleus and the overall size of the cell. Several lines of evidence suggest that these correlations are causal; that is, that adding DNA leads to an increase in the size of the nucleus as well as the cell in total.[13] First, there are numerous cases of organisms that have recently doubled their DNA content by doubling all of their chromosomes (and hence, DNA). The cells of these so-called polyploid organisms are larger than the cells of their ancestors who have less DNA per cell. Second, in the cases where within-species variation for DNA content exists, genome size and cell size are strongly correlated; for instance, cells from B-chromosome-containing individuals possess more DNA and tend to be bigger. Third, among subspecies and closely related species of leaf-eared mice in the genus *Phyllotis*, considerable variation for genome size and a strong correlation between genome size and cell size exist. Moreover, in crosses between species that differ in genome size, the hybrids are intermediate between the two species in the sizes of both their cells and their genomes. The mechanism through which increases in genome size lead to larger cells is still being debated, and these mechanisms may vary in the different groups of organisms.

■ Did Big Genomes Cause Salamanders to Evolve Simple Brains?

Some biologists who work with amphibians speculate that the large cells of these animals, owing to their large genome sizes, might constrain the complexity of their brains. More small cells than large cells can be packed into the same volume of space. If fewer cells can be packed in the brains of these species with large cells, does that mean that the large-genome, large-cell species have less complex brains?

Gerhard Roth, a neurobiologist at the University of Bremen in Germany, and David Wake, an evolutionary biologist at the University of California at Berkeley, have looked for associations between cell size (and thus genome size) and the complexity of one part of the brain in amphibians.[14] Roth and Wake chose to look at the region of the brain where most of the processing of visual information occurs both in amphibians and many other nonmammalian vertebrate groups, the area known as the optic tectum. As both frogs and amphibians are visual predators, the optic tectum is likely to be among the most important regions of the brain in these animals.

Despite the enormous sizes of genomes in some species of amphibians, the genome size of the ancestor to amphibians almost certainly was relatively modest for a vertebrate— not more than a couple billion nucleotides. Very large genome sizes appear to have independently evolved within amphibians—once in salamanders, and also in frogs. Although considerable variation in the genome sizes exists within both salamanders and frogs, the largest genomes of salamanders (100 billion nucleotides) are much larger than the largest of the frogs (20 billion nucleotides).

There is a strong tendency for frogs with small genomes and hence small cells to have more complex tectum morphologies than those with large genomes and large cells. This relationship holds up even when overall body and brain sizes are taken into account. In contrast, no correlation between brain size and either cell size or morphological complexity was observed in frogs.

Within the salamanders, the largest genomes and the largest cell volumes are found in the Plethodonitids. Sometimes referred to as the lungless salamanders, Plethodonitids in general develop less complex optic tecta compared with other salamanders. Within the Plethodonitids, salamanders belonging to the Bolitoglossini tribe have the largest genome sizes, approaching 100 billion. These salamanders invariably have optic tecta that are in the least complex categories. Bolitoglossini salamanders, which are terrestrial and lay eggs on land, occur mainly in the New World, specifically in Western North America, Central America, and South America.

Unlike frogs, salamanders show a correlation between brain size and cell size; the species with larger brains tend to have larger cells. When corrections are made for cell size, salamanders with larger brains develop more complex optic tecta. Cell size is negatively correlated with morphological complexity of the tecta.

The ancestral salamanders had smaller genomes and much more complex optic tecta than what is currently found in the Bolitoglossini. So it appears that the direction evolution took in this group was toward less sophisticated brain structures. This is a reminder that not all evolution is progressive or leads to more complex structures. Has this structural change in the brain been reflected in behavior or any other characteristic? That seems likely; Bolitoglossini salamanders adopted an ambush strategy for catching prey whereas their ancestors were active predators. The question, however, is which came first? Did the change in the behavior lead to reduced selection on the brain (specifically the optic tectum), which in turn allowed for genomes to get much larger? Or did the change in genome size lead to changes in the brain, which then led to changes in the way these salamanders catch their prey? No one knows.

Genome size is associated with modes of development in salamanders. Some species of plethodonitid salamanders undergo full metamorphosis, going through a distinct larval stage between egg and adult. Other species can undergo metamorphosis or hatch as miniature adults, depending on the circumstances. These are said to exhibit facultative metamorphosis. Genomes tend to be smaller in salamanders that undergo full metamorphosis than in the species whose metamorphosis is facultative. Salamanders that never undergo metamorphosis have the largest genomes of all.

Development in frogs also appears to be affected by genome size. *Bufo calamita, Rana temporana*, and other species of frogs that breed in temporary ponds are able to accelerate metamorphosis in response to the danger of desiccation. Genome size is lower in these species with accelerated metamorphosis than in their relatives, which do not breed in temporary ponds.[15]

This relationship between metamorphosis and small genome size is found not only in amphibians. Indeed, the common feature of insects with exceptionally large genomes (for instance, grasshoppers and crickets, cockroaches, walking sticks, dragonflies) is that they do not undergo complete metamorphosis with larval and pupal stages. These insects, instead, hatch from eggs as nymphs that look like smaller versions of the adults. There is no known insect with full metamorphosis whose genome is larger than two billion nucleotides.

■ Pufferfish and Shrunken Genomes

In contrast with most vertebrates, genomes of certain species of pufferfish are small and compact. Their smaller size coupled with less repetitive DNA makes genome sequencing easier; and for this reason, the genome of the smooth pufferfish *(Fugu rubripes)* was among the first of vertebrates to be sequenced. A draft of the smooth pufferfish genome sequence was published in 2002, the year after publication of the draft of the human genome.

Elizabeth Brainerd, then at the University of Massachusetts at Amherst, and her colleagues examined the relationship between genome size and morphology in pufferfish and their relatives.[16] Several interesting anatomical features characterize the order of fish within which pufferfish are classified (the Tetraodontiformes). This group of fish lacks ribs and the fins that would attach to their pelvises. They also evolved fewer vertebrae. These characteristics, however, do not appear associated with decreased genome size because within the group, genome size is not correlated with loss of anatomical parts; species of Tetraodontiformes fish with small genomes have about as many vertebrae as the ones with larger genomes.

In a subsequent study, Dan Neafsey and Stephen Palumbi at Stanford compared the smooth pufferfish *(Fugu)* to the related spiny pufferfish, whose genome is about twice as large.[17] They found less repetitive DNA in the smooth pufferfish genome than in that of the spiny pufferfish. For the most part, the patterns of insertions and deletions were similar in dead retrotransposons in these two species, despite their different genome sizes. One major difference was observed; fewer large insertions occurred in the smaller genome of *Fugu* than in that of the spiny pufferfish. Although we know that the excision of DNA occurs more rapidly in the smooth pufferfish than in most vertebrates, we still do not know the selective forces behind the reduction in genome size in these fish.

■ Effective Population Size and a Neutral Model for the Evolution of Genome Size

Michael Lynch at Indiana University recently published several articles on the provocative theme that the size and the complexity of genomes in eukaryotes did not arise because of positive selection. Instead, Lynch argues that the large, complex genomes of eukaryotes and the still larger and more complex genomes of multicellular organisms arose as a pathological consequence of the small population sizes of these organisms. Although selection is involved in shaping genome size and complexity in Lynch's model, it is the efficacy of negative selection, not positive selection, that is the primary determinant of properties of genomes.

Lynch begins with the intuitive and well-supported observation that organism size and population sizes are inversely correlated: species of small organisms (in general) exist in populations that are large, and vice versa. The range of population sizes is quite extensive—effective population sizes of most bacteria easily exceed a billion, whereas mammals almost never have effective population sizes above a million (and usually in the thousands). Note that census population sizes and effective population sizes need not be similar, as the human census population size is more than a hundred thousand times as large as the human effective population size (see chapter 3).

Lynch first turned to the presence and abundance of introns, the spacers in between the expressed portions of the genes. Ever since the discovery of interrupted genomes in the late 1970s, biologists have been puzzled by the patterns of intron abundance.[18] Most mammalian genes are broken up by a sizeable number of introns. It is not unusual for a gene to have tens of introns. Many introns in mammalian genes are quite long—much longer than the expressed parts of the genes (the exons). The sum total of intron DNA in humans is 20% of the total DNA, more than ten times the amount that is coding DNA. Although introns are abundant in insects, they are usually smaller and fewer in number per gene. Single-celled eukaryotes contain fewer introns still. Introns are virtually absent in bacteria and other prokaryotes.

Almost as soon as introns were discovered, a heated debate arose over their origin. Did introns arise before the split of bacteria and eukaryotes, and were lost in bacteria? Advocates of this view, the "introns early" position, claim that bacteria lost their introns because there was selective pressure on these cells to streamline their genomes. Another camp ("introns late") claims that introns appeared relatively late in the history of life—and clearly after eukaryotes split off from bacteria. The "introns late" advocates claim that bacteria never had introns to lose. The "introns early" and "introns late" positions actually represent two ends of a continuum, and several nuances exist within the debate, which are largely beyond the scope of this book.

Lynch's argument is that although introns may acquire functions that are beneficial for the organisms that harbor them, each intron begins by causing an extremely small detriment to fitness. Because effective population sizes are very high in most prokaryotes and many single-celled eukaryotes, natural selection is very efficient. Even though their fitness costs are extremely minute, introns are weeded out of microbes with large population sizes. In contrast, effective population sizes are modest in most multicellular organisms; selection is far less efficient here. Lynch proposes that introns are ever so slightly deleterious because the existence of the intron increases the likelihood that mutations in the genetic variant will lead to a nonfunctioning variant. With "back-of-the-envelope" calculations, Lynch estimates that new introns reduce fitness on the order of one part in 10 million.

In a population with an effective population size of one *billion*, natural selection would be efficient at removing genetic variants that reduce fitness by one in 10 million. Natural selection would not be efficient at removing such variants in a population with an effective population size of a *million;* such variants would be effectively neutral. Bacteria often have effective population sizes on the order of a billion, invertebrates have population sizes on the order of a few hundred thousand to a few million, and vertebrates typically have population sizes in the tens of thousands (or less). Lynch concluded:

> The results of this study suggest that the simple structure of genes in microbes relative to higher eukaryotes may have little to do with selective constraints (genomic streamlining in microbes) or adaptive requirements (the expansion of cellular complexity in multicellular eukaryotes). Rather, the genomes of species with habitually large population sizes may simply be immunized from the spread of introns by the power of secondary mutation.[19]

With computer scientist John Conery, Lynch extended his argument to the presence of transposable elements.[20] Based on the assumption that most excess DNA is slightly deleterious, they argued that there should be an inverse relation between effective population size and genome size. Species with large effective population sizes should have smaller genomes because natural selection is more efficient when effective population sizes are large than when they are small.

There is a general negative correlation between effective population size and genome size, consistent with the predictions of Lynch and Conery. This result, however, comes with several caveats. Effective population size is difficult to measure, especially in microbes. Another potential concern is the way in which the species in the data set were chosen; the strength of the conclusion reached could depend on the choice of species.

The most serious caveat is that the study may confuse cause with effect. Bigger genomes can lead to bigger body sizes, which in turn can lead to lower population sizes. From this argument, we would expect the pattern of an inverse correlation between genome size and effective population size, but not for the reasons advanced by Lynch and Conery.

■ Testing the Lynch Model

What is of importance in Lynch's model is the long-term effective population size, not short-term fluctuations. Thus, if a species had a recent genetic bottleneck, its genome size should not be affected. Moreover, several other forces influence genome size. Effective population size is just part of the answer, but it does appear to explain part of the genome size enigma.

One apparent challenge to Lynch and Conery's model involved ciliates, single-celled eukaryotes that are often rather large in size. Many are visible to the naked eye, and some approach a tenth of an inch in length. This group is named for their most distinguishing feature, hair-like projections known as cilia. The ciliates also possess a number of other interesting properties, including having two different types of nuclei.[21]

One ciliate, *Tetrahymena thermophila*, has a rather large genome—about 200 million nucleotides, similar in size to the *Drosophila* genome and not drastically smaller than the pufferfish genome—but had been thought to have a large effective population size. Such a pattern of high effective population size and large genome size, if frequent, would be evidence against Lynch's model. The basis for the claim that *Tetrahymena* thermophila had a large effective population size was the discovery of high levels of genetic diversity at a single genetic locus, *Ser*H.

Laura Katz at Smith College and her colleagues examined five other genes from this ciliate species.[22] In each, they found far less genetic diversity than previous workers had at *Ser*H. These results led Katz and her colleagues to conclude that *Tetrahymena* thermophila actually has a rather modest effective population size—comparable to that of most insects. They argue that *Ser*H has probably been under balancing selection (chapter 5) and suggest that *Ser*H, which codes for a protein found on the surface of the cell, might play a role similar to genes that produce antigens in vertebrate immune systems. Such vertebrate genes—like those of the major histocompatibility complex—are frequently the targets of balancing selection. The new, lower estimate of effective population for *Tetrahymena* as a whole is consistent with the prediction of Lynch and Conery.

Soojin Yi and Todd Streelman at the Georgia Institute of Technology tested the hypothesis of Lynch and Conery in ray-finned fish, also known as bony fish.[23] This group arose from a single common ancestor approximately 425 million years ago and has evolved considerable range in genome size. The genome of the smooth pufferfish, which we met earlier in the chapter, contains just 370 million nucleotides whereas that of short-nosed sturgeon has over 6 billion nucleotides (twice the size of the human genome).

It has been long known that the effective population sizes of most species of freshwater fish are lower than those of most marine fish species. This is probably the case because freshwater species, being confined to rivers and lakes and other freshwater sources of water, tend to have much lower geographical distributions than marine species, many of which have a whole ocean—or more than one ocean—to swim in.

On average, freshwater species do indeed have statistically significant larger genomes than do marine species. In the former, the average genome size is 1.5 billion nucleotides, whereas in the latter, the average genome size is just one billion nucleotides. Thus, there is support at least in ray-finned fish for the model proposed by Lynch and Conery.

■ Genome Size and Extinction

Genome size appears to affect the survival of species. Alexander Vinogradov at the Russian Academy of Sciences showed that plants with big genomes are at greater risk of extinction than plants with small genomes.[24] Vinogradov's criteria for high risk of extinction was whether the plant was already extinct or was listed at any level of endangered or threatened according to the official "red list."

For fish, genome size is higher in species at high risk for extinction than in those that are at low risk. This pattern, however, arises because of differences in extinction risk and genome size among the largest taxonomic groups. Within these large groups, no pattern emerges. Sharks, rays, and other cartilaginous fish have larger genomes and a greater proportion of high-risk species than do ray-finned fish. Looking within the cartilaginous fish, no relationship exists between extinction risk and genome size. Within the ray-finned fish, the genomes of sturgeons are larger than those of other ray-finned fish, and high-risk species are more frequent in the former than the latter. Excluding the sturgeons, there is no relationship between genome size and extinction risk. It is interesting that no relationship has been seen between genome size and extinction risk in mammals, regardless of the categories within mammals used in the analyses.

Vinogradov argues that conservation biologists should consider genome size data as a tool. He states, "It may be convenient to use genome size (which can now be determined quite easily and accurately using flow cytometry) as an additional predictor of potential conservation status of a given species."[25]

■ Coda

Darwin found grandeur in "a tangled bank," a complicated natural world that arose from the interplay of several fundamental laws of nature: the struggle for existence, inheritance, and natural selection, among others. He saw that the "war of nature" led to the production of "endless forms, most beautiful and most wonderful." Although Darwin had no knowledge of the complicated world that exists within genomes, Darwin's intellectual descendants find that similar Darwinian principles apply to that world. The ability of genomic parasites to invade is probably a consequence of random processes overwhelming the ineffective natural selection found in species with large bodies but relatively small effective population sizes. Moreover, the very complexity of genes and genomes seen in vertebrates and land plants appears to be the result of a similar process. Investigations of the overall size and other properties of the genome may have practical importance, but their extent is yet unknown. The importance of studying the natural histories of genomes and genes, like the importance of studying the natural histories of organisms, lies not in what such studies can do to save the world but rather in the discovery itself.

■ Recommended Reading

General information on the evolution of genome size can be found in the following references.

Burt, A., and R. Trivers. 2006. *Genes in Conflict: The Biology of Selfish Genetic Elements*. Harvard University Press.

Gregory, T. R. 2001. Coincidence, coevolution, or causation? DNA content, cell size, and the C-value enigma. *Biological Review* 76: 65–101.

Gregory, T. R. 2002. Genome size and developmental complexity. *Genetica* 115: 131–146.

Gregory, T. R. 2005. Synergy between sequence and size in large-scale genomics. *Nature Reviews Genetics* 6: 699–708.

Preface

1. See for instance, Friedman, 2000.
2. http://www.google.com/corporate/timeline.html
3. The pace of the sequencing of the human genome was accelerated when Craig Venter, then head of the private company Celera Genomics, challenged the publicly funded consortium that led the effort in the Human Genome Project. Celera and the public consortium acted in part as rival competitors and in part as collaborators during the late 1990s. Each group published a draft sequence (about 90% complete) of the human genome in February 2001; Celera's appeared in *Science* (Venter et al., 2001) and the consortium's appeared in *Nature* (The International Human Genome Sequencing Consortium, 2001). In the April 24, 2003 issue of *Nature*, the National Human Genome Research Institute announced that mapping of the sequence was essentially completed, and it outlined future challenges (Collins et al., 2003).

Chapter 1

1. Bailey's statement and his reflections on Baby Fae in December 1999 can be found at http://www.llu.edu/news/today/dec299/mc.htm.
2. Antolin and Herbers, 2001, p. 2384. Other accounts of the Baby Fae case include Jonasson and Hardy (1985) and Stoller (1990).
3. A leading evolutionary biology textbook (Futuyma, 1998, p. 768) defines homology as "possession by two or more species of a trait derived, with or without modification, from their common ancestor." Different species can have similar traits for reasons other than sharing common ancestry; biologists call such similarity, homoplasy.

4. "Bush unveils $7.1 billion plan to prepare for flu pandemic." http://www.cnn.com/2005/HEALTH/conditions/11/01/us.flu.plan/ (accessed April 2006).

5. Harris, 2006. World Health Organization's list of confirmed H5N1 cases as of 4 May 2006: http://www.who.int/csr/disease/avian_influenza/country/ cases_table_2006_05_04/ en/index.html. "From the Front Lines," 2006.

6. Kuiken et al., 2006. These authors also outline other factors that currently hinder avian flu in the human population. For instance, the avian and mammalian flu strains differ in the amino acid at position 627 in the polymerase (the enzyme involved in the replication of DNA): the avian flu has glutamic acid and the mammalian flu has lysine. This difference appears to limit the avian flu from replicating in mammalian cells, because avian viruses that have been engineered to have lysine at this position are more virulent. One isolate of the H5N1 avian flu taken from China in 2005 ominously contains the amino acid lysine at that position.

7. Specter, 2005.

8. Dr. Fauci is quoted in Nickerson (2005).

9. Taubembuger et al., 2005. General information about the 1918–19 flu pandemic can be found in Barry (2005), who noted that the flu was called the "Spanish flu" because the newspapers providing the most graphic coverage of the pandemic in Europe were from Spain. This country was the largest nonaligned country during World War I, and as such did not impose wartime censorship on its media.

10. Zhou et al. (1999) provide an early example of rapid evolution in the bird flu.

11. Kuiken et al., 2006, p. 397.

12. Facts on the rise of antibiotic resistance and its impact on hospital infections can be found at http://www.niaid.nih.gov/factsheets/antimicro.htm (accessed May 2006).

13. Senator William Frist, a medical doctor and the former U.S. Senate majority leader, said in 1999 in a hearing of the Subcommittee on Public Health regarding the problem of antimicrobial resistance, "It is estimated that $30 billion is spent on the cumulative effects of antimicrobial resistance each year." See http://www.acponline.org/ear/factsfigs.htm (accessed May 2006). In the seven years since Senator Frist's statement the problem has only become worse.

14. Palumbi, 2001. In addition to material on the evolution of resistance in HIV, Palumbi provides several more case examples of the use of evolutionary principles to counter the evolution of resistance in bacteria to antibiotics and in insect pests to pesticides.

15. Mouse Genome Sequencing Consortium, 2002. There is a one-to-one correspondence for mouse and human genes for approximately 80% of the genome. Only 1% of human genes lack a mouse counterpart, and vice versa. The difference between these two figures mainly arises from cases in which humans and mice differ with respect to how many copies of a gene they have.

16. Reviewed in Rand et al. (2004).

17. The most extensive survey of variation in the human genome to date is from the Hap Map Project (The International HapMap Consortium, 2005).

18. Rose, 1991.

19. Eichner et al., 2002. A slightly modified version of Eichner et al. can be found at http://www.cdc.gov/genomics/hugenet/reviews/APOEcardio.htm.

20. Price et al., 2004.

21. Although the original use and exact wording of Haldane's quip remain subject to debate, it is clear that Haldane himself was fond (perhaps inordinately so) of repeating it. Gould (1995) discusses the history of this anecdote.

22. Ray's Tierra system is described in Kelly (1995), Yedid and Bell (2001), and Johnson (2001).
23. Metzker et al., 2002.
24. Information about the McVeigh case was taken from http://www.cnn.com/US/ 9706/17/mcveigh.overview (accessed May 2006).
25. Kliman and Johnson, 2005.

Chapter 2

1. Information about the Kansas decision can be found in Slevin (2005) and Wilgoren (2005).
2. Governor Sebelius is quoted in Wilgoren (2005): "If we're going to continue to bring high-tech jobs to Kansas and move our state forward, we need to strengthen science standards, not weaken them.... Stronger public schools ought to be the mission of the Board of Education, and it's time they got down to the real business of strengthening Kansas schools."
3. Mervis, 2005.
4. The complete text of the Kitzmiller trial decision can be found at this site: http://www.pamd.uscourts.gov/kitzmiller/decision.htm. The decision includes the full text of the disclaimer that was read to the Dover students:

    The Pennsylvania Academic Standards require students to learn about Darwin's Theory of Evolution and eventually to take a standardized test of which evolution is a part.

    Because Darwin's Theory is a theory, it continues to be tested as new evidence is discovered. The Theory is not a fact. Gaps in the Theory exist for which there is no evidence. A theory is defined as a well-tested explanation that unifies a broad range of observations. Intelligent Design is an explanation of the origin of life that differs from Darwin's view. The reference book, *Of Pandas and People*, is available for students who might be interested in gaining an understanding of what Intelligent Design actually involves.

    With respect to any theory, students are encouraged to keep an open mind. The school leaves the discussion of the Origins of Life to individual students and their families. As a Standards-driven district, class instruction focuses upon preparing students to achieve proficiency on Standards-based assessments. (pp. 1–2)
5. Behe, 1996. Behe's research career prior to publication of *Darwin's Black Box* involved the structure and biochemical properties of a form of DNA known as Z DNA.
6. Ibid, p. 5.
7. Larson (1997) provides a detailed account on William Jennings Bryan and his role at the Scopes trial.
8. Behe, 1996, p. 5.
9. Ibid, p. 39.
10. The "argument from design" has a long history, going back to at least the Greek philosophers some 24 centuries before Paley.
11. Darwin, 1859, p. 217.
12. General information about the evolution of the eye is taken from Dawkins (1996).
13. Nilsson and Pelger, 1994.

14. Miller (1999) provides the basic model for the evolution of the cascade involved in blood clotting. Miller's book also provides many other examples of how what appears as irreducible complexity could arise from Darwinian processes. Miller discusses more about the biochemical cascade involved in blood clotting here: http://www.millerandlevine.com/km/evol/DI/clot/Clotting.html (Accessed April 2006). See also Doolittle (1997) for further details.

15. Jiang and Doolittle, 2003.

16. The discussion between Miller and Behe occurred on April 23, 2002. It is part 7 of a transcript from the American Museum of Natural History, and can be found here: http://www.ncseweb.org/resources/articles/7819_07_dr_michael_behe_dr_10_31_2002.asp (Accessed May 2006).

17. Pomiankowski et al., 2004. One central consideration in Pomiankowski's model is that a particular variant of a gene may have different fitness in males and females; for instance, it may be beneficial in males but slightly deleterious in females. If the positive effect that that variant has in males is greater than its negative effect in females, then that variant will be favored. The negative effect of that variant in females, however, sets up the possibility of further changes at different genes, another main feature of Pomiankowski's model.

18. Harris, 1990.

19. The quote by Reagan and more exposition on the difference between fact and theory in science can be found in Johnson (2007).

20. Asimov made this remark to the National Center Against Censorship (NCAC) in 1980. http://www.ncac.org/censorship_news/20030305~cn064~Sticky_Solution_-_Glue_the_Pages.cfm (Accessed June 2006).

21. Mooney (2005, p. 167) describes the early days of the Discovery Institute.

22. The mission statement of the Discovery Institute can be found at its homepage: http://www.discovery.org/about.php (Accessed June 2006).

23. Forrest and Gross, 2004, p. 10. See also Grant (2004).

24. The "Wedge Document" can be read here: http://www.antievolution.org/features/wedge.html (Accessed June 2006). The Center for the Renewal of Science and Culture is the previous name of the Center for Science and Culture.

25. Detailed in Forrest and Gross, 2004.

26. Dembski 1999, p. 206

27. Ibid, p. 207.

28. Ibid, p. 210.

29. Krauthammer, 2005.

30. Will, 2005. The Jefferson quote is from *Notes on Virginia* (1782).

31. Larson, 1997.

32. The views of the Episcopal Church on science and evolution can be found at the following Web site: http://www.episcopalchurch.org/19021_58398_ENG_HTM.htm (Accessed June 2006).

33. Schönborn, 2005.

34. "Communion and Stewardship: Human Persons Created in the Image of God," plenary sessions held in Rome 2000–2002, published July 2004, paragraph 63.

35. Schönborn's speech can be found at http://www.bringyou.to/apologetics/p91.htm (Accessed May 2006).

36. The clergy letter was signed by over 10,000 clergy members as of November 2005. It can be found here: http://www.uwosh.edu/colleges/cols/religion_science_collaboration.htm (Accessed June 2006).

37. Peters and Hewlett, 2003. See also: http://www.plts.edu/docs/EvolutionBrief2.pdf. See also Henry, 1998.
38. See in particular, Shanks, 2004, p. 139: "The central stumbling blocks for intelligent design theory actually have little to do with pernicious materialistic philosophies alleged to be held by its opponents. The central stumbling blocks are all evidential in nature. The accusation that scientists reject intelligent design theory because they are in the sway of materialistic or naturalistic philosophy is part of a smoke-and-mirrors strategy to cover this sad reality from public scrutiny."
39. The quote on Fisher's religious views comes from Orr, 2005, p. 52.
40. Dobzhansky, 1945, p. 74.
41. Gilkey is quoted in Numbers, 1992, p. xv.
42. Goodstein, 2005. Interestingly, the Rev. Warren Eshbach was the spokesperson for "Dover Cares," the group of pro-science candidates. His son, Robert, was one of those elected to the board in 2005.
43. Kitzmiller decision, p. 136. See note 4.
44. Ibid, p. 64.
45. Ibid, p. 49.
46. Heath, 2006.
47. Lerner, 2006, p. 131.

Interlude I

1. A plethora of Darwin biographies exist. These biographies and other accounts detail how the observations made on his voyage on the *Beagle* led Darwin to formulate his theory of evolution. My favorite among these is the first volume of Janet Browne's two-volume biography on Darwin (Browne, 1995).
2. Jonathan Weiner's *The Beak of the Finch* (1994) is a wonderful account of the Grants' work on examining evolution by natural selection in Darwin's finches. Freeman and Herron (2001) discuss the conditions for natural selection in the context of the Grants' work on Darwin's finches.
3. Gilchrist et al., 2001.
4. Grant, 2005; Grant and Wiseman, 2002.
5. Darwin was greatly influenced by the works of Thomas Malthus, an eighteenth-century economist who pessimistically warned of dangers of famine and pestilence because human population growth had a tendency to outstrip growth in food supply. Malthusian thought also shaped the ideas of Alfred Wallace, the co-discoverer of natural selection.
6. Despite his brilliance in many areas, Darwin was not much of a mathematician. In fact, the first edition of *The Origin of Species* contains a small error in the calculation of the expected number of elephants given unchecked reproduction. This error, which was corrected in subsequent editions, did not invalidate Darwin's central message that even organisms with very slow rates of reproduction could not have their populations grow unchecked forever.

Chapter 3

1. The examples of rates of evolutionary change in fibrinopeptides and histone H4 are taken from Kimura (1983), the best source for information about the early

development of the neutral theory. Although several of the chapters are highly mathematical, others are quite accessible to non-mathematically inclined readers. Kimura (1979) presents a shorter and less technical version of his theory.

2. The figure of 1.2 billion years since the common ancestor of plants and animals means that there has been a total of 2.4 billion years of evolution because the two lineages have each been evolving separately. The estimate of 1.2 billion years may be off by hundreds of millions of years in either direction, but it is certainly accurate to a factor of two in either direction.
3. Crow (1987) and Ohta (1996) present biographical information on Kimura.
4. Kimura, 1968, 1983.
5. Lewontin, 1974. See also http://hrst.mit.edu/hrs/evolution/public/transcripts/origins_transcript.html.
6. King and Jukes, 1969. Crow (2000, p.955) recounted the history of the paper by King and Jukes, saying: "They submitted a manuscript to *Science*, only to have it turned down. One reviewer said it was obviously true and therefore trivial; the other said that it was obviously wrong. King and Jukes appealed, and the second time it was accepted."
7. Mayr, 1963, p. 207. Mayr, who lived past 100, would come to change his mind regarding the importance of neutral genetic variants in light of the evidence amassed by molecular evolutionary biologists.
8. The differences in views between Coyne et al. (1997) and Wade and Goodnight (1998) demonstrate the current state of Wright's theories.
9. Kimura, 1983, pp. 178–183.
10. See Lewontin (1974) for information on the struggle to measure genetic variability in natural populations, and the results from the early days of protein electrophoresis.
11. See chapter 8 of Wright (1969) and chapter 2 of Wright (1978) for an overview of effective population size.
12. Mayr, 1991, p. 152.

Chapter 4

1. Bargelloni et al. (1998) present information about the specialized adaptations of hemoglobins in Antarctic dragonfish and other Antarctic fish.
2. Hemoglobin's response to acidity is called the Bohr effect.
3. Bishop et al., 2000.
4. Kreitman, 1983.
5. McDonald and Kreitman, 1991.
6. Ibid, p. 653.
7. In fact, some scientists were initially skeptical of the validity of the McDonald-Kreitman because it seemed just too simple to be correct. The following year, the population geneticists Stanley Sawyer and Daniel Hartl (1992) published a theoretical paper showing that the McDonald-Kreitman test was indeed valid.
8. Maynard Smith and Haigh, 1974.
9. Hilton et al., 1994. The "dot chromosome" is the fourth chromosome of *Drosophila melanogaster*.
10. Pritchard and Przeworski, 2001.
11. Schlenke and Begun, 2005.
12. The International HapMap Consortium, 2005.
13. Bierne and Eyre-Walker, 2004.

Chapter 5

1. For general information about sickle-cell anemia, see http://www.sicklecelldisease.org/about_scd/ (Accessed June 2006).
2. Pauling et al., 1949.
3. Hager, 1995, p. 334.
4. Ingram, 1957.
5. General information on malaria is from http://malaria.who.int/ (Accessed June 2006).
6. Williams et al., 2005.
7. Fairhurst et al., 2005.
8. Lewontin, 1974, p. 199.
9. Kerem et al., 1989.
10. O'Brien (2003) provides basic information on the link between variation in the chemokine receptor *CCR5* gene and HIV resistance, as well as a first-hand account on his lab's work in this area.
11. Bamshad et al., 2002; Johnson, 2002b.
12. Lewontin et al., 1984; Rosenberg et al., 2002.
13. Hedrick et al., 1991.
14. Meyer and Thomson (2001) provide a detailed review about the evidence for balancing selection operating at HLA. They also discuss the possible mechanisms by which balancing selection has operated.

Interlude II

1. Darwin, 1871.
2. Huxley, 1861. The entire Huxley files can be found at the following site: http://alepho.clarku.edu/huxley/ (Accessed June 2006).
3. Sutikna et al., 2004; Morwood et al., 2004.
4. Dawkins (2004, pp. 90–99) discusses the timing of bipedalism and various hypotheses for why it may have occurred.
5. Diamond, 1992, p. 32.

Chapter 6

1. Haley, 1976.
2. Pearson, 2003; Editorial, 1998.
3. Harmon, 2005.
4. Due to their bacterial origins and their unusual mode of transmission, mitochondria have several unusual properties; see Rand et al. (2004) for further details. Most animals, including vertebrates and the vast majority of insects, have maternal transmission of mitochondria. There are some exceptions. For instance, individuals of many species of mussels receive their mitochondria from both their mother and their father (Piganeau et al. 2004).
5. Mullis, 1990.
6. Cann et al., 1987.
7. Templeton, 1993.
8. Vigilant et al., 1991.
9. Ingman et al., 2000, p. 709.

10. In the late 1990s, geneticists discovered that the transmission of mammalian mitochondrial DNA is not as strictly maternal as once thought. Not only do documented cases of fathers transmitting some mitochondria to their offspring exist (Schwartz and Vissing, 2002), but it appears that the mtDNA of the father and that of the mother can recombine (Kraytsberg et al., 2004). The incidence of paternal transmission and recombination appears to be extremely low, but as often happens in evolutionary biology, events with extremely low frequencies can have substantial effects on the conclusions that we can draw. This finding does not invalidate the geographical aspects of the work, but it is possible that this minute level of paternal transmission and recombination could affect the dating of Eve. Although (as of 2006) there is no consensus as to how much effect minute levels of recombination would have on the dating, my educated guess is this effect is sufficiently minor that other factors would render it null.

11. Sykes, 2001.

12. Foster et al., 1998.

13. Jones (2002) and Wells (2002) provide general information about the evolution of the Y chromosome and the tracing of its history in humans.

14. Dorit et al., 1995.

15. Underhill et al., 2000.

16. Seielstad et al. 1999.

17. Takahata et al., 2001.

18. Templeton, 2002.

19. Dawkins, 2004, pp. 40–47.

20. Chang, 1999, p. 1003.

Chapter 7

1. Shreeve (1995) provides general information on Neanderthals.

2. Churchill is quoted in Augarde (1991, p. 55).

3. Though officially retired, Brace remains active in physical anthropology. He still considers Neanderthals to be our direct ancestors. See Brace (1995).

4. Harvati et al. 2004.

5. Ramirez Rozzi and Bermudez De Castro, 2004, p. 98.

6. Duarte et al., 1999.

7. Tattersall and Schwartz, 1999.

8. Standard practice in the application of PCR is to use a DNA polymerase from bacteria adapted to high temperature. Such polymerases, unlike most proteins, will maintain their function even after exposure to very high temperatures.

9. Krings et al., 1997.

10. Hoss, 2000, p. 453.

11. Ovchinnikov, 2000.

12. Caramelli, 2003, p. 6593.

13. Serre et al., 2004.

14. Currant and Excoffier, 2004, p. 2266.

15. Lalueza-Fox et al., 2005.

16. Some anthropologists have claimed that some Neanderthal morphological features were present at low frequency in early modern humans in Europe. For instance, almost all Neanderthal skulls have a slight depression at the base. Reltheford (2003, p. 98) notes that this feature, known as the suprainiac fossa, is rare in living

Europeans but is found in 39% the post-Neanderthal moderns. Such a finding is consistent with a model in which modern humans and Neanderthals shared some genes, and then modern humans outcompeted the Neanderthals. This result isn't strong evidence in favor of such a model to the exclusion of a model wherein Neanderthals did not share many genes with our forebears. First, it is not certain whether the feature seen in the "post-Neanderthal moderns" that is called the suprainiac fossa is indeed the same feature in the Neanderthals; that is, the homology is in doubt. Second, even if we are dealing with homologous features, the genetic bases of the suprainiac fossa and other features are not well known. It is possible that some of these features may have a strong environmental component, which could explain their gradual disappearance even if Neanderthals and early modern humans didn't exchange genes.

17. Coyne and Orr, 2004.
18. Mellars, 2006.
19. Green et al., 2006. Pääbo's team used a new technique called pyrosequencing to ascertain the sequence of approximately one million nucleotides scattered among different chromosomes of the Neanderthal genome. Their findings confirm the tentative conclusions based on the mitochondrial DNA data; modern humans and Neanderthals last shared a common ancestor that existed about half a million years ago. These data also suggest that the effective population size of this ancestral species was only a few thousand; the lineage that modern humans and Neanderthals share has had a low effective population size for several hundred thousand years.

Chapter 8

1. General background about chimpanzees, bonobos, and the differences between them is taken from de Waal, 2005.
2. de Waal, 2005, p. 7.
3. de Waal, 1995.
4. Parish, 1994.
5. A small minority of anthropologists had favored placing humans and orangutans as each other's closest relative. The DNA evidence soundly rejects this hypothesized pairing.
6. Freeman and Herron (2001, pp. 552–557) provide more detail on gene trees versus species trees in light of the human-chimpanzee-gorilla phylogeny.
7. Ruvolo, 1997.
8. Wise et al., 1997.
9. Kaessman et al., 1999.
10. Wong and Hey, 2005.
11. The Chimpanzee Sequencing and Analysis Consortium, 2005. The white paper: http://www.genome.gov/Pages/Research/Sequencing/SeqProposals/Chimp_Geno mel_editted.pdf (Accessed June 2006).
12. Marks, 2002.

Chapter 9

1. See Vargha-Khadem et al. (1998, 2005) and Lai et al. (2001) for general information on the language defect found in the KE family, as well as on the *FOXP2* gene.

2. In a PET (positron emission tomography) scan, a sugar is tagged with a radioactive isotope with a short half-life. The use of that sugar by various parts of the brain can be visualized, and that provides a measure of the activity.

3. Functional magnetic resonance imaging (fMRI), which relies on the fact that oxygenated and deoxygenated hemoglobin have different magnetic properties, traces blood flow through sections of the brain.

4. Haesler et al., 2004.

5. Enard et al., 2002.

6. See Walton (2002) for the CNN story.

7. Jackendoff, 1999.

8. Evans et al., 2004a; see also Gilbert et al. (2005) for a general introduction to the work by Lahn's group.

9. Evans et al., 2005. Technically, because there is variation within it, what I referred to as a haplotype is actually a haplogroup.

10. Evans et al., 2004b.

11. Merkel-Bolorov et al., 2005.

12. Gould, 1996. There is a modest correlation of brain size and general intelligence, but that correlation could be confounded by environmental factors, including socioeconomic status and family background. Within families, the correlation between brain size and general intelligence is, at best, weak. Moreover, considerable controversy still exists regarding exactly what IQ tests actually measure.

13. Chou et al., 2002.

14. Ibid.

15. Martin et al., 2005.

16. Clark et al., 2003.

17. Nielsen et al., 2005.

18. Eberhard, 1996.

19. Programmed cell death is sometimes called apoptosis, after a Greek word meaning "falling off" in the sense that leaves fall off trees.

20. Gilbert et al., 2005.

Chapter 10

1. Pennisi, 2004.

2. The extinct Australian aboriginal language Damin, which is used today only in ceremonial rituals, is the only known non-African click language. Damin probably acquired clicks from exposure to click languages. One click is often used by English speakers—a dental click, made by pulling the tongue away from the teeth to make a sharp sibilant sound. It is not used in any words, but, when repeated, it indicates disapproval and is sometimes transcribed as "tut tut" or "tsk tsk."

3. See McWhorter, 2001, page 35–36. McWhorter misidentifies the grandsons of Charlemagne as the sons of Charlemagne.

4. Cavalli-Sforza, 2000, p. 164.

5. Jones is quoted in Pennock (1999, p. 134).

6. Pennock, p. 127.

7. Burling, 2005, p. 2.

8. Cavalli-Sforza and Cavalli-Sforza, 1995.

9. Croft (2001) and Wade (2005) provide information on Greenberg. A corrected version of Croft's essay can be found at http://lings.ln.man.ac.uk/Info/staff/WAC/Papers/JHGobit.pdf.
10. Cavalli-Sforza and Cavalli-Sforza, 1995, p. 174.
11. Knight et al., 2003.

Chapter 11

1. Darwin, 1868 (1998), p. 15.
2. Coppinger and Coppinger, 2001.
3. Ibid, p. 74.
4. Coppinger and Coppinger (2001) and Truit (1999) describe the Belyaev experiment.
5. Truit (1999), p. 163.
6. Vila et al., 1997.
7. Savolainen et al., 2002.
8. Parker et al., 2004. Ostrander and her team have since moved to the National Human Genome Research Institute at the National Institutes of Health.
9. Lindblad-Toh et al., 2005.
10. Matsuoka et al., 2002; Fukunaga et al., 2005; Johnson, 2002a.
11. Wang et al., 1999.
12. Destro-Bisol et al., 2004.

Chapter 12

1. Five hundred and six books by Isaac Asimov are listed at http://www.asimovonline.com/oldsite/asimov_catalogue.html (Accessed May 2006).
2. Gregory, 2002.
3. Ibid. See also Burt and Trivers (2006, chapter 7).
4. Orgel and Crick, 1980, p. 605.
5. Keller, 1983.
6. Burt and Trivers, 2006, chapter 7.
7. Gregory, 2005.
8. Burt and Trivers, 2006, pp. 249–255.
9. The estimate for the extent that P element load affects fitness in *Drosophila melanogaster* depends on the relationship between fitness and the number of individual P elements in a fly. If the relationship is linear (having two P elements are twice as bad as having one), then the load is about 20%. If instead the relationship is non-linear such that having two P elements are more than twice as bad having one, then the overall load is closer to 10%. For more information, see Burt and Trivers (2006, p. 253) and references therein.
10. Petrov et al., 1996.
11. Petrov et al., 2000.
12. Petrov and Hartl, 2000.
13. Gregory, 2001.
14. Roth et al., 1994.
15. Gregory, 2002.
16. Brainerd et al., 2001.
17. Neafsey and Palumbi, 2003.

18. Lewin, 1997, chapter 22.
19. Lynch, 2002, p. 6123.
20. Lynch and Conery, 2003.
21. The two nuclei divide the two major roles of DNA: replication of genetic information to the next generation, and expression of that genetic information via the transcription of DNA information into RNA information and the translation of RNA information into protein information. A small nucleus contains the DNA that is transmitted to new generations; this is known as the micronucleus. The other nucleus, much larger in size, is involved in the transcription and translation of DNA information into protein information; this is the macronucleus.
22. Katz et al., 2006.
23. Yi and Streelman, 2005.
24. Vinogradov (2003, 2004) has the most comprehensive treatment on the relationship between genome size and extinction rate.
25. Vinogradov, 2004, p. 1704.

Adaptation. A phenotypic characteristic that increases the likelihood of viability or reproductive output of individuals bearing it in comparison with individuals that lack the characteristic.

Amino acid. The fundamental unit of proteins. Proteins are long chains of amino acids; the identity of a protein is determined mainly by the sequence of its amino acids.

Balancing selection. The condition wherein selection actively maintains two or more genetic variants in a population or species. Balancing selection can result from a number of different underlying mechanisms. One type of balancing selection occurs when the heterozygote has a higher fitness than do either of the two homozygotes. Balancing selection can also arise from rare variants being favored over common ones or geographic variation in the selective regime.

Codon. Three sequential nucleotides of coding DNA that specify an amino acid.

DNA (Deoxyribonucleic acid). The genetic material; a long double-stranded molecule whose sequence carries genetic information. The information in DNA can perform two roles: (1) transmission to the next generation, and (2) specification the types of proteins to be produced.

Dominant. In a heterozygote, the variant that is expressed. The recessive variant is not expressed.

Effective population size (Ne). A measure of the extent genetic drift will affect frequencies of genetic variants in a population. It is roughly equal to the number of breeding individuals of an ideal population that remains constant in size and has equal numbers of males and females. Populations with high effective population sizes will experience less genetic drift than those with lower Ne.

Exon. Part of gene expressed into proteins (cf. Intron).

Fixation. The situation when one genetic variant at a site reaches frequency of 1, and thus eliminates all other variants.

Gene. Geneticists employ several different definitions for the term, "gene." One commonly used definition is a sequence of nucleotides in a strand of DNA that codes for a protein.

Genetic drift. The random evolutionary process that alters frequencies of genetic variants within a population due to sampling error every generation. Such sampling error can occur during the process of the production of gametes (eggs and sperm), the fusion of gametes to make zygotes, and chance variation in individuals surviving and/or reproducing.

Genome. The complete DNA sequence in the cells of a given organism.

Genotype. Set of genes that an individual organism possesses. It can also refer to the particular genetic variant that an individual harbors at a specific site along the genome.

Great Leap Forward. According to Jared Diamond, a period roughly 50,000 years ago when human cultural change accelerated.

Haplotype. Short for "haploid genotype." a haplotype is a set of closely linked genetic variants that are inherited as a unit.

Heterozygote. An individual that possesses two different variants at a given nucleotide or gene (cf. homozygote).

Hitchhiking. The increase in frequency of a genetic variant at one site due to being genetically linked with a favored variant at another site.

Homozygote. An individual that possesses two of the same variants at a given nucleotide or gene (cf. heterozygote).

Hypothesis. A provisional statement that attempts to explain observations and/or phenomena of the natural world. Sets of hypotheses that have been confirmed via tests against other observations become theories.

Intelligent design (ID). The proposition that some or all aspects of living organisms are too complex to arise by natural processes and therefore must have been designed by an "Intelligent Designer."

Intron. Part of gene that is not expressed into proteins. It is spliced out from the RNA (cf exon).

Irreducible complex. A system is considered irreducibly complex if it is "composed of several well-matched, interacting parts that contribute to the basic function, wherein the removal of any one of the parts causes the system to effectively stop functioning" (Behe, 1996, p. 39).

Lineage. A branch of the evolutionary tree; a lineage can consist of a single species or two or more species that share a common ancestor.

Linkage disequilibrium (LD). Correlation between genetic variants at different, but linked, genetic sites.

McDonald-Kreitman test. A test that enables one to detect whether a gene region has been subject to positive selection by observing patterns of polymorphism and divergence at replacement and silent sites.

Mitochondria. Organelles in cells involved in aerobic respiration that originally came from free-living bacteria-like organisms.

Mitochondrial DNA (mtDNA). DNA from the mitochondria, a small organelle in the cell. In most animals, mtDNA is a relatively small molecule and is transmitted only through females.

Mitochondrial Eve (mtEve). Woman that was the most recent common ancestor of all mitochondrial DNA variants in a population (cf. Y-chromosome Adam).

Mutation. Any change in the DNA sequence of an individual.

Natural selection. Differential survival or reproductive ability of genotypes due to phenotypic differences among the genotypes.

Negative selection. Selection that weeds out deleterious genetic variants.

Neutral theory (of molecular evolution). Developed by Motoo Kimura, the neutral theory explains expected patterns of polymorphism and divergence seen in DNA and amino-sequences if only mutation, genetic drift, and negative selection are operating.

Nucleotide. Fundamental unit of DNA; the identity of a DNA molecule or fragment thereof is determined by its sequence of nucleotides.

PCR (polymerase chain reaction). Three-part biochemical reaction that results in a doubling of specified DNA molecules each cycle of the reaction.

Parsimony. Philosophical principle that one should prefer the least complex hypothesis that explains a phenomenon.

Phenotype. Refers to all observable or measurable characteristics of an individual organism. Phenotypic traits include morphological, behavioral, physiological, and biochemical ones.

Phylogeny. Evolutionary history of a group.

Pleiotopy. Variation at one gene having multiple phenotypic effects.

Polymorphism. Literally multiple forms; variation at a gene or nucleotide in a population or species. Also a genetic variant.

Positive selection. Selection that results in an advantageous genetic variant increasing in frequency in the population.

Random genetic drift. See genetic drift.

Recombination. The shuffling of genes due to exchange of chromosomes during gamete formation.

Recessive. In a heterozygote, the variant that is not expressed. The dominant variant is expressed.

Replacement change. Change in nucleotide sequence of DNA that results in change to amino-acid sequence (cf. silent change).

Reproductive isolation. The inability of individuals from one population to mate with and/or produce viable and fertile offspring with individuals from another population. Reproductive isolation is a continuum; populations that are sufficiently reproductively isolated from each another are considered species by the most commonly used definition of species.

Selective sweep. The increase in frequency of an advantageous genetic variant due to positive selection. The region of the genome surrounding a selective sweep is often characterized by low levels of genetic variation and high levels of linkage disequilibrium.

Silent change. A change in a nucleotide of protein-coding DNA that does not change the amino acid (cf. replacement change).

Single nucleotide polymorphism (SNP). Variation in a population or species at a single nucleotide site.

Statistically significant. A result that is sufficiently unlikely to be due to chance as determined by a statistical test. Typically, the level of significance is 5%.

Substitution. Fixation of a once-rare mutation in a population or species.

Theory. A set of statements that (1) are interconnected and internally consistent, (2) are based on observable and circumstantial evidence, (3) explain a wide variety of observations, and (4) have been tested by reference to the observable facts. Theories are broader than hypotheses; indeed, theories are sets of hypotheses that have been confirmed.

Transcription. The specification of RNA genetic information from DNA information.

Translation. The specification of protein genetic information from RNA information.

Transposable element. Also known as a "jumping gene" and a "mobile element," this is a piece of DNA that can be copied and inserted into other regions of the genome.

Y-chromosome Adam. Man that was the most recent common ancestor of all Y-chromosome variants in a population (cf. Mitochondrial Eve).

References

Antolin, M. F., and J. M. Herbers. 2001. Evolution's struggle for existence in America's public schools. *Evolution* 55: 2379–2388.

Augarde, T., 1991. *The Oxford Dictionary of Modern Quotations.* Oxford University Press.

Bamshad, M. J., S. Mummidi, E. Gonzalez, S. S. Ahuja, D. M. Dunn, W. S. Watkins, S. Wooding, A. C. Stone, L. B. Jorde, R. B. Weiss, and S. K. Ahuja. 2002. A strong signature of balancing selection in the 5' *cis*-regulatory region of *CCR5 Proceedings of the National Academy of Sciences (USA)* 99: 10539–10544.

Bargelloni, L., S. Marcato, and T. Patarnello. 1998. Antarctic fish hemoglobins: Evidence for adaptive evolution at subzero temperature. *Proceedings of the National Academy of Sciences (USA)* 95: 8670–8675.

Barry, J. M., 2005. *The Great Influenza: The Epic Story of the Deadliest Plague in History.* Extended Paperback edition. Penguin Books.

Behe, M. J., 1996. *Darwin's Black Box: The Biochemical Challenge to Evolution.* Free Press.

Bierne, N., and A. Eyre-Walker. 2004. The genomic rate of adaptive amino acid substitution in *Drosophila. Molecular Biology and Evolution* 21: 1350–1360.

Bishop, J. G., A. M. Dean, and T. Mitchell-Olds. 2000. Rapid evolution in plant chitinases: Molecular targets of selection in plant-pathogen coevolution. *Proceedings of the National Academy of Sciences (USA)* 97: 5322–5327.

Brace, C. L., 1995. Biocultural interaction and the mechanism of mosaic evolution in the emergence of "modern" morphology. *American Anthropologist* 97: 711–723.

Brainerd, E. L., S. S. Slutz, E. K. Hall, and R. W. Phillis. 2001. Patterns of genome size in Tetradontiform fishes. *Evolution* 55: 2363–2368.

Browne, J., 1995. *Voyaging.* Knopf.

Burling, R., 2005. *The Talking Ape: How Language Evolved*. Oxford University Press.

Burt, A., and R. Trivers. 2006. *Genes in Conflict: The Biology of Selfish Genetic Elements*. Harvard University Press.

Cann, R. L., M. Stoneking, and A. C. Wilson. 1987. Mitochondrial DNA and human evolution. *Nature* 325: 31–36.

Caramelli, D., C. Lalueza-Fox, C. Vernesi, M. Lari, A. Casoli, F. Mallegni, B. Chiarelli, L. Dupanloup, J. Bertranpetit, G. Barbujani, and G. Bertorelle. 2003. Evidence for a genetic discontinuity between Neandertals and 24,000-year-old anatomically modern Europeans. *Proceedings of the National Academy of Sciences (USA)* 100: 6593–6597.

Cavalli-Sforza, L. L. 2000. *Genes, People, and Languages*. University of California Press.

Cavalli-Sforza, L. L., and F. Cavalli-Sforza. 1995. *Great Human Diasporas: The History of Diversity and Evolution*. Helix Books.

Chang, J. T., 1999. Recent common ancestors of all present-day individuals. *Advances in Applied Probability* 31: 1002–1026.

Chou, H. H., T. Hayakawa, S. Diaz, M. Krings, E. Indriati, M. Leakey, S. Pääbo, Y. Satta, N. Takahata, and A. Varki. 2002. Inactivation of CMP-*N*-acetylneuraminic acid hydroxylase occurred prior to brain expansion during human evolution. *Proceedings of the National Academy of Sciences (USA)* 99: 11736–11741.

Clark, A. G., S. Glanowski, R. Nielsen, P. D. Thomas, A. Kejariwal, M. A. Todd, D. M. Tanenbaum, D. Civello, F. Lu, B. Murphy, S. Ferriera, G. Wang, X. G. Zheng, T. J. White, J. J. Sninsky, M. D. Adams, and M. Cargill. 2003. Inferring nonneutral evolution from human-chimp-mouse orthologous gene trios. *Science* 302: 1960–1963.

Collins, F. S., E. D. Green, A. E. Guttmacher, and M. S. Guyer on behalf of the US National Human Genome Research Institute. 2003. A vision for the future of genomics research. *Nature* 422: 835–847.

Coppinger, R., and L. Coppinger. 2001. *Dogs: A Startling New Understanding of Canine Origin, Behavior, and Evolution*. Scribner.

Coyne, J. A., N. H. Barton, and M. Turelli. 1997. Perspective: A critique of Sewall Wright's shifting balance theory of evolution. *Evolution* 51: 643–671.

Coyne, J. A., and H. A. Orr. 2004. *Speciation*. Sinauer Associates, Inc.

Croft, W., 2001. Joseph Harold Greenberg. *Language* 77: 815–830.

Crow, J. F., 1987. Twenty-Five years ago in *Genetics*: Motoo Kimura and molecular evolution. *Genetics* 116: 183–184.

Crow, J. F., 2000. Thomas H. Jukes (1906–1999). *Genetics* 154: 955–956.

Currant, M., and L. Excoffier. 2004. Modern humans did not admix with Neanderthals during their range expansion into Europe. *Public Library of Science Biology* 2: 2264–2274.

Darwin, C., 1859. *The Origin of Species*. John Murray.

Darwin, C., 1868. *The Variation of Animals and Plants Under Domestication*. Volume 1. (reprinted and edited by H. Ritvo, 1998). Johns Hopkins.

Darwin, C., 1871. *The Descent of Man and Selection in Relation to Sex*. John Murray

Dawkins, R., 1996. *Climbing Mount Improbable*. W. W. Norton and Company.

Dawkins, R., 2004. *The Ancestor's Tale: A Pilgrimage to the Dawn of Evolution*. Houghton Mifflin.

de Waal, F. B. M., 1995. Bonobo sex and society. *Scientific American* 272 (3): 82–88 (March).

de Waal, F. B.M., 2005. *Our Inner Ape: A Leading Primatologist Explains Why We Are Who We Are.* Riverhead.

Dembski, W.A., 1999. *Intelligent Design: The Bridge Between Science & Theology.* InterVarsity Press.

Destro-Bisol, G., F. Donati, V. Coia, I. Boschi, F. Verginelli, A. Caglià, S. Tofanelli, G. Spedini, and C. Capelli. 2004. Variation of female and male lineages in Sub-Saharan populations: the importance of sociocultural factors. *Molecular Biology and Evolution* 21: 1673–1682.

Diamond, J., 1992. *The Third Chimpanzee: The Evolution & Future of the Human Animal.* HarperCollins.

Diamond, J., 1997. *Guns, Germs, and Steel: The Fates of Human Societies.* W. W. Norton & Co.

Diamond, J., 2002. Evolution, consequences and future of plant and animal domestication. *Nature* 418: 700–707.

Dobzhansky, Th., 1945, Review of Marsh. *The American Naturalist* 79: 73–75.

Doolittle, R. L., 1997. A delicate balance. *Boston Review.* February/March 1997, pp. 28–29. Published on the web at http://www.bostonreview.net/br22.1/doolittle.html

Dorit, R. L., H. Akashi, and W. Gilbert. 1995. Absence of polymorphism at the ZFY Locus on the human Y Chromosome. *Science* 268: 1183–1185.

Duarte, C., J. Maurício, P. Pettitt, P. Souto, E. Trinkaus, H. van der Plicht, and J. Zilhão. 1999. The early Upper Paleolithic human skeleton from the Abrigo do Lagar Velho (Portugal) and modern human emergence in Iberia. *Proceedings of the National Academy of Sciences (USA)* 96: 7604–7609.

Eberhard, W. G., 1996. *Female Control: Sexual Selection by Cryptic Female Choice.* Princeton University Press.

Editorial. 1998. Genome vikings. *Nature Genetics* 20: 99–101.

Ehrlich, P. R., 2000. *Human Natures: Genes, Cultures, and the Human Prospect.* Penguin Books.

Eichner, J. E., S. T. Dunn, G. Perveen, D. M. Thompson, K. E. Stewart, and B. G. Stoehla. 2002. Apolipoprotein E polymorphism and cardiovascular Disease: A HuGE review. *American Journal of Epidemology* 155: 487–495.

Enard, W., M. Przeworski, S. E. Fisher, C. S. L. Lai, V. Wiebe, T. Kitano, A. P. Monaco, and S. Pääbo. 2002. Molecular evolution of *FOXP2*, a gene involved in speech and language. *Nature* 418: 869–872.

Evans, P. D., J. R. Anderson, E. J. Vallender, S. S. Choi, B. T. Lahn. 2004a. Reconstructing the evolutionary history of *microcephalin*, a gene controlling human brain size. *Human Molecular Genetics* 13: 1139–1145.

Evans, P. D., J. R. Anderson, E. J. Vallender, S. L. Gilbert, C. M. Malcom, S. Dorus, B. T. Lahn. 2004b. Adaptive evolution of *ASPM*, a major determinant of cerebral cortical size in humans. *Human Molecular Genetics* 13:489–494.

Evans, P. D., S. L. Gilbert, N. Mekel-Bobrov, E. J. Vallender, J. R. Anderson, L. M. Vaez-Azizi, S. A. Tishkoff, R. R. Hudson, and B. T. Lahn. 2005. *Microcephalin*, a gene regulating brain size, continues to evolve adaptively in humans. *Science* 309: 1717–1720.

Fairhurst, R. M., and 15 coauthors. 2005. Abnormal display of PfEMP-1 on erythrocytes carrying haemoglobin C may protect against malaria. *Nature* 435: 1117–1121.

Forrest, B., and P. R. Gross. 2004. *Creationism's Trojan Horse: The Wedge of Intellect Design.* Oxford University Press.

Foster, E. A., M. A. Jobling, P. G. Taylor, P. Donnelly, P. de Knijff, R. Mieremet, T. Zerjal, C. Tyler-Smith. 1998. Jefferson fathered slave's last child. *Nature* 396: 27–28.

Freeman, S., and J. Herron. 2001. *Evolutionary Analysis*. Prentice Hall.

Friedman, T. L., 2000. *The Lexus and the Olive Tree: Understanding Globalization*. Anchor Books.

From the Front Lines. 2006. *Nature* 440: 726–727.

Fukunaga, K., J. Hill, Y. Vigouroux, Y. Matsuoka, J. Sanchez, K. J. Liu, E. S. Buckler, and J. Doebley. 2005. Genetic diversity and population structure of teosinte. *Genetics* 169: 22412254.

Futuyma, D. J., 1998. *Evolutionary Biology* (third edition). Sinauer Associates, Inc.

Gilbert, S. L., W. B. Dobyns, and B. T. Lahn. 2005. Genetic links between brain development and brain evolution. *Nature Review Genetics* 6: 581–590.

Gilchrist, G. W., R. B. Huey, and L. Serra. 2001. Rapid evolution of wing size clines in *Drosophila subobscura*. *Genetica* 112/113: 273–286.

Goodstein, L., 2005. A decisive election in a town roiled over intelligent design. *New York Times*, November 10, 2005, p. A1.

Gould, S. J., 1995. *Dinosaur in the Haystack*. Harmony Books.

Gould, S. J., 1996. *The Mismeasure of Man* (second edition). W. W. Norton & Co.

Grant, B., 2004. Intentional Deception: Intelligent Design Creationism. Skeptic http://www.skeptic.com/eskeptic/04-06-01.html

Grant, B. S., 2005. Industrial melanism. In *Encyclopedia of Life Sciences*. John Wiley & Sons.

Grant, B. S., and L. L. Wiseman. 2002. Recent history of melanism in American peppered moths. *Journal of Heredity* 93: 86–90.

Green, R. E., J. Krause, S. E. Ptak, A. W. Briggs, M. T. Ronan, J. F. Simons, L. Du, M. Egholm, J. M. Rothberg, M. Paunovic, and S. Pääbo. 2006. Analysis of one million base pairs of Neanderthal DNA. *Nature* 444: 330–336.

Gregory, R. T., 2001. Coincidence, coevolution, or causation? DNA content, cell size, and the C-value enigma. *Biological Review* 76: 65–101.

Gregory, R. T., 2002. Genome size and developmental complexity. *Genetica* 115: 131–146.

Gregory, R. T., 2005. Synergy between sequence and size in large-scale genomics. *Nature Reviews Genetics* 6: 699–708.

Hager, T., 1995. *Force of Nature: The Life of Linus Pauling*. Simon and Schuster.

Haesler, S., K. Wada, A. Nshdejan, E. E. Morrisey, T. Lints, E. D. Jarvis, and C. Scharff. 2004. *FoxP2* Expression in Avian Vocal Learners and Non-Learners. *The Journal of Neuroscience* 24: 3164–3175.

Haley, A., 1976. *Roots*. Doubleday.

Harmon, A., 2005. Blacks pin hope on DNA to fill slavery's gaps in family trees. *New York Times*, July 25, 2005.

Harris, S., 1990. *You Want Proof—I'll Give You Proof: Sidney Harris Laughs at Science*. W. H. Freeman & Company.

Harris, G., 2006. States welcome flu plan but say they need federal money. *New York Times*, May 4, 2006, p. A20.

Harvati, K., S. R. Frost, and K. P. McNulty. 2004. Neanderthal taxonomy reconsidered: Implications of 3D primate models of intra- and interspecific differences. *Proceedings of the National Academy of Sciences (USA)* 101: 1147–1152.

Heath, E., 2006. Evolution after Dover. *BioScience* 56: 300.

Hedrick, P. W., W. Klitz, W. P. Robinson, M. K. Kuhner, and G. Thomson. 1991. Population genetics of HLA. In *Evolution at the Molecular Level*. (R. K. Selander, A. G. Clark, and T. S. Whitman, eds., pp. 248–271). Sinauer Press.

Henry, G. C., 1998. *Christianity and the Images of Science*. Smith & Helwys.

Hilton, H., R. M. Kliman, and J. Hey. 1994. Using hitchhiking genes to study adaptation and divergence during speciation in the *Drosophila melanogaster* species complex. *Evolution* 48: 1900–1913.

Hoss, M., 2000. Ancient DNA: Neanderthal population genetics. *Nature* 404: 453–454.

Huxley. T. H., 1861. On relations of man to the lower animals. http://alepho.clarku.edu/huxley/CE7/RelM-l-a.html

Ingman, M., H. Kaessmann, S. Pääbo, and U. Gyllensten. 2000. Mitochondrial genome variation and the origin of modern humans. *Nature* 408: 708–713.

Ingram, V. M., 1957. Gene mutations in human hemoglobin: the chemical difference between normal and sickle hemoglobin. *Nature* 180: 326–328.

Jackendoff, R., 1999. Possible stages in the evolution of the language capacity. *Trends in Cognitive Science* 3: 272–279.

Jiang, Y., and R. L. Doolittle. 2003. The evolution of vertebrate blood coagulation as viewed from a comparison of puffer fish and sea squirt genomes. *Proceedings of the National Academy of Sciences (USA)* 100: 7527–7532.

Johnson, N. A., 2001. Do Tierran programs dream of Darwinian dynamics? *Trends in Genetics* 17: 491–492.

Johnson, N. A., 2002a. One origin for maize. *Trends in Genetics* 18: 344.

Johnson, N. A., 2002b. Strong balancing selection in the regulatory region of *CCR5*. *Trends in Genetics* 18: 501.

Johnson, N. A., 2007. Is Evolution only a theory?: Scientific methodologies and evolutionary biology. In Petto A.J., Godfrey L.R., eds., *Scientists Confront Intelligent Design and Creationism*, 2nd edition. W.W. Norton (in press).

Jonasson, O., and M. A. Hardy. 1985. The case of Baby Fae. *Journal of the American Medical Association* 254:3358–3359.

Jones, S., 2002. *Y: The Descent of Man*. Little, Brown.

Judson, O., 2005. Evolution is in the air. *New York Times*. Op-ed. November 6, 2005.

Katz, L. A., O. Snoeyenbos-West, and F. M. Doerder. 2006. Patterns of protein evolution in *Tetrahymena thermophila*: Implications for estimates of effective population size. *Molecular Biology and Evolution* 23: 608–614. .

Keller, E. F., 1983. *A Feeling for the Organism*. W. H. Freeman and Company.

Kelly, K., 1995. *Out of Control: The New Biology of Machines, Social Systems, and the Economic World*. Perseus Books Group.

Kerem, B., J. M. Rommens, J. A. Buchanan, D. Markiewicz, T. K. Cox, A. Chakravarti, M. Buchwald, and L.C. Tsui. 1989. Identification of the cystic fibrosis gene: Genetic analysis. *Science* 245: 1073–1080.

Kimura, M., 1968. Evolutionary rate at the molecular level. *Nature* 217: 624–626.

Kimura, M., 1979. The neutral theory of molecular evolution. *Scientific American* 241(11): 98–126.

Kimura, M., 1983. *The Neutral Theory of Molecular Evolution*. Cambridge University Press.

King, J. L., and T. H. Jukes, 1969. Non-Darwinian evolution. *Science* 164: 788–798.

Kliman, R. M., and N. A. Johnson. 2005. What (and why) every undergraduate should know about evolution. *BioScience* 55: 926–927.

Knight, A., P. A. Underhill, H. M. Mortensen, L. A. Zhivotovsky, A. A. Lin, B. M. Henn, D. Louis, and J. L. Mountain. 2003. African Y chromosome and mtDNA divergence provides insight into the history of click languages. *Current Biology* 13: 464–473.

Krauthammer, C., 2005. Phony theory, false conflict: "Intelligent Design" foolishly pits evolution against faith. *Washington Post*, November 18, 2005, p. A23.

Kraytsberg, Y., M. Schwartz, T. A. Brown, K. Ebralidse, W. S. Kunz, D. A. Clayton, J. Vissing, K. Khrapko. 2004. Recombination of human mitochondrial DNA. *Science* 304: 981.

Kreitman, M., 1983. Nucleotide polymorphism at the alcohol dehydrogenase locus of *Drosophila melanogaster. Nature* 304: 412–417.

Krings, M., A. Stone, R. W. Schmitz, H. Krainitzki, M. Stoneking, and S. Pääbo. 1997. Neandertal DNA sequences and the origin of modern humans. *Cell* 90: 19–30.

Kuiken, T., E. C. Holmes, J. McCauley, G. F. Rimmelzwaan, C. S. Williams, B. T. Grenfell. 2006. Host species barriers to Influenza virus infections. *Science* 312: 394–397.

Lai, C.S.L., S. E. Fisher, J. A. Hurst, F. Vargha-Khadem, and A. P. Menance. 2001. A forkhead-domain gene is mutated in a severe speech and language disorder. *Nature* 413:519–523.

Lalueza-Fox, C., M. L. Sampietro, D. Caramelli, Y. Puder, M. Lari, F. Calafell, C. Martínez-Maza, M. Bastir, J. Fortea, M. de la Rasilla, J. Bertranpetit, and A. Rosas. 2005. Neandertal evolutionary genetics: Mitochondrial DNA data from the Iberian Peninsula. *Molecular Biology and Evolution* 22: 1077–1081.

Larson, E. J., 1997. *Summer for the Gods: The Scopes Trial and America's Continuing Debate Over Science and Religion.* Harvard University Press.

Lerner, M., 2006. *The Left Hand of God.* HarperSanFranciso.

Lewin, B., 1997. *Genes VI.* Oxford University Press.

Lewontin, R. C., 1974. *The Genetic Basis of Evolutionary Change.* Columbia University Press

Lewontin, R. C., S. Rose, and L. J. Kamin. 1984. *Not in Our Genes: Biology, Ideology and Human Nature.* Pantheon.

Lindblad-Toh, K., C. M. Wade, T. S. Mikkelsen, E. K. Karlsson, D. B. Jaffe, and many others. 2005. Genome sequence, comparative analysis and haplotype structure of the domestic dog. *Nature* 438: 803–819.

Lynch, M., 2002. Intron evolution as a population-genetic process. *Proceedings of the National Academy of Science (USA)* 99: 6118–6123.

Lynch, M., 2006. The origins of eukaryotic gene structure. *Molecular Biology and Evolution* 23: 450–468.

Lynch, M., and J. S. Conery. 2003. The origins of genome complexity. *Science* 302: 1401–1404.

Marks, J. 2002. *What It Means to be 98% Chimpanzee.* University of California Press.

Marmom, W., 1977. Haley's Rx: Talk, write, reunite. *Time* 109: 72–73. February 14, 1977.

Martin, M. J., J. C. Rayner, P. Gagneux, J. W. Barnwell, and A. Varki. 2005. Evolution of human—chimpanzee differences in malaria susceptibility: Relationship to human genetic loss of N-glycolylneuraminic acid. *Proceedings of the National Academy of Sciences (USA)* 102: 12819–12824.

Matsuoka, Y., Y. Vigouroux, M. M. Goodman, J. Sanchez G., E. Buckler, and J. Doebely. 2002. A single domestication for maize shown by multilocus microsatellite genotyping. *Proceedings of the National Academy of Sciences* 99: 6080–6084.

Maynard Smith, J., and J. Haigh 1974. The hitch-hiking effects of a favorable gene. *Genetical Research* 23: 23–25.

Mayr, E., 1963. *Animal Species and Evolution*. Harvard University Press.

Mayr, E., 1991. *One Long Argument*. Harvard University Press.

McDonald, J. H., and M. Kreitman. 1991. Adaptive protein evolution at the *Adh* locus in *Drosophila*. *Nature* 351: 652–654.

McWhorter, J., 2001. *The Power of Babel: A Natural History of Language*. W. H. Freeman.

Mellars, P., 2006. A new radiocarbon revolution and the dispersal of modern humans in Eurasia. *Nature* 439: 931–935.

Merkel-Bolorov, M., S. L. Gilbert, P. D. Evans, E. J. Vallender, J. R. Anderson, R. R. Hudson, S. A Tishkoff, and B. T. Lahn. 2005. Ongoing Adaptive Evolution of *ASPM*, a Brain Size Determinant in *Homo sapiens*. *Science* 309: 1720–1722.

Mervis, J., 2005. Evolution: Dover teachers want no part of Intelligent-Design statement. *Science* 307: 505.

Metzker, M. L., D. P. Mindell, X. M. Liu, R. G. Ptak, R. A. Gibbs, and D. M. Hillis. 2002. Molecular evidence of HIV-1 transmission in a criminal case. *Proceedings of the National Academy of Sciences (USA)* 99: 14292–14297.

Meyer, D., and G. Thomson. 2001. How selection shapes variation of the human major histocompatibility complex: a review. *Annals of Human Genetics* 65: 1–26.

Miller, G., 2000. *The Mating Mind: How Sexual Choice Shaped the Evolution of Human Nature*. Doubleday.

Miller, K. R., 1999. *Finding Darwin's God: A Scientist's Search for Common Ground Between God and Evolution*. Cliff Street Books.

Mooney, C., 2005. *The Republican War on Science*. Basic Books.

Morwood, M. J., and 14 coauthors. 2004. Archaeology and age of a new hominin from Flores in eastern Indonesia. *Nature* 431: 1087–1091.

Mouse Genome Sequencing Consortium. 2002. Initial sequencing and comparative analysis of the mouse genome. *Nature* 420: 520–562.

Mullis, K., 1990. The unusual origin of the polymerase chain reaction. *Scientific American* 262 (4): 56–65 (April 1990).

Neafsey, D. E., and S. R. Palumbi. 2003. Genome size evolution in pufferfish: A comparative analysis of diodontid and tetraodontid pufferfish genomes. *Genome Research* 13: 821–830.

Nickerson, C., 2005. Avian flu strain confirmed in Romania. *Boston Globe*. October 16, 2005, p. A1. http://www.boston.com/news/nation/articles/2005/10/16/avian_ flu_strain_confirmed_in_romania/

Nielsen, R., C. Bustamante, A. G. Clark, S. Glanowski, T. B. Sackton, M. J. Hubisz, A. Fledel-Alon, D. M. Tanenbaum, D. Civello, T. J. White, J. J. Sninsky, M. D. Adams, and M. Cargill. 2005. A scan for positively selected genes in the genomes of humans and chimpanzees. *Public Library of Science Biology* 3: e170.

Nilsson, D. E., and S. Pelger. 1994. A pessimistic estimate of the time required for an eye to evolve. *Proceedings of the Royal Society of London B.* 256: 53–58.

Numbers, R. L. 1992. *The Creationists: The Evolution of Scientific Creationism*. University of California Press.

O'Brien, S. J., 2003. *Tears of the Cheetah: The Genetic Secrets of Our Animal Ancestors*. Thomas Dunne Books, St. Martin Griffin.

Ohta, T., 1996. Motoo Kimura. *Annual Review of Genetics* 30: 1–5.

Orgel, L. E., and F. H. C. Crick. 1980. Selfish DNA: The ultimate parasite. Nature 284: 604–607.

Orr, H. A., 2005. Devolution: Why Intelligent Design Isn't. *The New Yorker* 81 (15): 40–52. May 30, 2005. On-line at http://www.newyorker.com/fact/content/articles/050530fa_fact

Ovchinnikov, I. V., A. Gotherstrom, G. P. Romanova, V. M. Kharitonov, K. Liden, and W. Goodwin. 2000. Molecular analysis of Neanderthal DNA from the northern Caucasus. *Nature* 404:490–493.

Palumbi, S. R., 2001. *The Evolution Explosion: How Humans Cause Rapid Evolutionary Change*. W. W. Norton & Company.

Parker, H. K., L., V. Kim, N. B. Sutter, S. Carlson, T. D. Lorentzen, T. B. Malek, G. S. Johnson, H. B. DeFrance, E. A. Ostrander, and L. B. Kruglyak. 2004. Genetic structure of the purebred domestic dog. *Science* 304: 1160–1164.

Parish, A. P., 1994. Sex and food control in the uncommon chimpanzee: How bonobo females overcame a phylogenetic legacy of male dominance. *Ethology and Sociobiology* 15: 157–194.

Pauling, L., H. A. Itano, S. J. Singer, and I. C. Wells, 1949. Sickle cell anemia: A molecular disease. *Science* 110: 543–548.

Pearson, H., 2003. Profile: Kári Stefánsson. *Nature Medicine* 9: 1099.

Pennisi, E., 2004. The first language? *Science* 303: 1319–1320.

Pennock, R. T., 1999. *Tower of Babel: The Evidence Against the New Creationism*. The MIT Press.

Peters, T., and M. Hewlett. 2003. *Evolution from Creation to New Creation: Conflict, Conversation, and Convergence*. Abingdon Press

Petrov, D.A., E. R. Lozovskaya, and D. L. Hartl. 1996. High intrinsic rate of DNA loss in *Drosophila*. *Nature* 384, 346–349.

Petrov, D. A., and D. L. Hartl. 2000. Pseudogene evolution and natural selection for a compact genome. *The Journal of Heredity* 91: 221–227.

Petrov, D. A., T. A. Sangster, J. S. Johnston, D. L. Hartl, and K. L. Shaw. 2000. Evidence for DNA loss as a determinant of genome size. *Science* 287: 1060–1062.

Piganeau, G., M. Gardner, and A. Eyre-Walker. 2004. A broad survey of recombination in animal mitochondria. *Molecular Biology and Evolution* 21: 2319–2325.

Pomiankowski, A., R. Nöthiger, and A. Wilkins. 2004. The evolution of the *Drosophila* sex-determination pathway. *Genetics* 166: 1761–1773.

Price, A. L., E. Eskin, and P. A. Pevzner. 2004. Whole-genome analysis of *Alu* repeat elements reveals complex evolutionary history. *Genome Research* 14: 2245–2252.

Pritchard, J. K., and M. Przeworski. 2001. Linkage disequilibrium in humans: Models and data. *American Journal of Human Genetics* 69: 1–14.

Rand, D. M., R. A, Haney, and A. J. Frey. 2004. Cytonuclear coevolution: The genomics of cooperation. *Trends in Ecology and Evolution* 19: 647–655.

Relethford, J. H., 2003. *Reflections of Our Past: How Human History Is Revealed in Our Genes*. Westview Press.

Rose, M. R. 1991. *Evolutionary Biology of Aging*. Oxford University Press.

Rosenberg, N. A., J. K. Pritchard, J. L. Weber, H. M. Cann, K. K. Kidd, L. A. Zhivotovsky, and M. W. Feldman. 2002. Genetic structure of human populations. *Science* 298: 2381–2385.

Roth, G., J. Blanke, and D. B. Wake, 1994. Cell size predicts morphological complexity in the brains of frogs and salamanders. *Proceedings of the National Academy of Science (USA)* 91: 4796–4800.

Ramirez Rozzi, F. V., and J. M. Bermudez De Castro. 2004. Surprisingly rapid growth in Neanderthals. *Nature* 428: 936–939.

Ruvolo, M. 1997. Molecular phylogeny of the homonids: Inferences from multiple independent DNA data sets. *Molecular Biology and Evolution* 14: 248–265.

Sagan, C., and A. Druyan. 1992. *Shadows of Forgotten Ancestors: A Search for Who We Are.* Random House.

Savolainen, P., Y. P. Zhang, J. Luo, J. Lundeberg, T. Leitner. 2002. Genetic evidence for an East Asian origin of domestic dogs. *Science* 298: 1610–1613.

Sawyer, S. A., and D. L. Hartl. 1992. Population genetics of polymorphism and divergence. *Genetics* 132: 1161–1176.

Schlenke, T. A., and D. J. Begun. 2005. Linkage disequilibrium and recent selection at three immunity receptor loci in *Drosophila simulans*. *Genetics* 169: 2013–2022.

Schönborn, C., 2005. Finding design in Nature. *The New York Times.* Op-ed. July 9, 2005.

Schwartz, M., and J. Vissing. 2002. Paternal inheritance of mitochondrial DNA. *New England J. of Medicine* 347: 576–580.

Scott, E. C., 2005. *Evolution vs. Creationism: An Introduction.* (paperback ed.) University of California Press.

Seielstad, M., E. Bekele, M. Ibrahim, A.Touré, and M. Traoré. 1999. A view of modern human origins from Y chromosome microsatellite variation. *Genome Research* 9: 558–567.

Serre, D., A. Langaney, M. Chech, M. Nicola-Teschler, M. Paunovic, P. Mennecier, M. Hofreiter, G. Possnert, and S. Pääbo. 2004. No evidence of neandertal mtDNA contribution to early modern humans. *PLOS Biology* 2: 313–317.

Shanks, N., 2004. *God, the Devil, and Darwin: A Critique of Intelligent Design Theory.* Oxford University Press.

Shreeve, J., 1995. *The Neandertal Enigma: Solving the Mystery of Modern Human Origins.* W. Morrow and Co.

Slevin, P., 2005. Kansas education board first to back "Intelligent Design." *Washington Post*, November 9, 2005, p. A1.

Smith, B. D., 1998. *The Emergence of Agriculture* (paperback edition). Scientific American Library.

Specter, M., 2005. Nature's bioterrorist: Is there any way to prevent a deadly avian-flu pandemic? *The New Yorker*, 49–61. (February 28, 2005)

Stoller, K. P., 1990. Baby Fae: The unlearned lesson. *Perspectives on Medical Research*, Volume 2. Retrieved from http://www.curedisease.com/Perspectives/vol_2_1990/BabyFae.html

Stringer, C., 2002. New perspectives on the Neanderthals. *Evolutionary Anthropology* 11: 58–59, Suppl. 1.

Shreeve, J., 1995. *The Neandertal Enigma: Solving the Mystery of Modern Human Origins.* W. W. Morrow and Co.

Sutikna, T., M. J. Morwood, R. P. SoejonoJatmiko, E. Wayhu Saptomo, and R. Awe Due. 2004. A new small-bodied hominin from the Late Pleistocene of Flores, Indonesia. *Nature* 431: 1055–1061.

Sykes, B., 2001. *The Seven Daughters of Eve: The Science That Reveals Our Genetic Ancestry.* W. W. Norton and Co.

Takahata, N., S. H. Lee, and Y. Satta. 2001. Testing multiregionality of modern human origins. *Molecular Biology and Evolution* 18: 172–183.

Tattersall, I., and J. M. Schwartz. 1999. Hominids and hybrids: The place of Neanderthals in human evolution. *Proceedings of the National Academy of Sciences (USA)* 96: 7117–7119.

Taubenburger, J. K., A. H. Reid, R. M. Lorens, R. Wang, G. Jin, and T. G. Fanning. 2005. Characteristics of the 1918 influenza virus polymerase genes. *Nature* 437: 889–893.

Templeton, A. R., 1993. The "Eve" hypothesis: a genetic critique and reanalysis. *American Anthropologist* 95:51–72.

Templeton, A. R., 2002. Out of Africa again and again. *Nature* 416: 45–51.

The Chimpanzee Sequencing and Analysis Consortium. 2005. Initial sequence of the chimpanzee genome and comparison with the human genome. *Nature* 437: 69–87.

The International HapMap Consortium. 2005. A haplotype map of the human geonome. *Nature* 437: 1299–1320.

The International Human Genome Sequencing Consortium. 2001. Initial sequencing and analysis of the human genome. *Nature* 409: 860–921.

Truit, L. M. 1999. Early canid domestication: The farm-fox experiment. *American Scientist* 87: 160–168.

Underhill, P. A., P. Shen, A. A. Lin, L. Jin, G. Passarino, W. H. Yang, E. Kauffman, B. Bonné-Tamir, J. Bertranpetit, P. Francalacci, M. Ibrahim, T. Jenkins, J. R. Kidd, S. Q. Mehdi, M. T. Seielstad, R. S. Wells, A. Piazza, R. W. Davis, M. W. Feldman, L. L. Cavalli-Sforza & P. J. Oefner. 2000. Y chromosome sequence variation and the history of human populations. *Nature Genetics* 26: 358–361.

Vargha-Khadem, F., D. G. Gadian, A. Copp, and M. Mishkin. 2005. *FOXP2* and the neuroanatomy of speech and language. *Nature Reviews Neuroscience* 6: 131–138.

Vargha-Khadem F., K. E. Watkins, C. J., Price, J. Ashburner, K. J. Alcock, A. Connelly, R. S. J. Frackowiak, K. J. Friston, M. E. Pembrey, M. Mishkin, D. G. Gadian, and R. E. Passingham. 1998. Neural basis of an inherited speech and language disorder. *Proceedings of the National Academy of Science (USA)* 95: 12695–12700.

Venter, J. C. (and numerous authors), 2001. The sequence of the human genome. *Science* 291: 1304–1351.

Vigilant, L., M. Stoneking, H. Harpending, K. Hawkes, and A. C. Wilson. 1991. African populations in the evolution of human mitochondrial DNA. *Science* 253: 1503–1507.

Vila, C., P. Savolainen, J. E. Maldonado, I. R. Amorim, J. E. Rice, R. L. Honeycutt, K. A. Crandall, J. Lundeberg, and R. K. Wayne. 1997. Multiple and ancient origins of the domestic dog. *Science* 276: 1687–1689.

Vinogradov, A. E. 2003. Selfish DNA is maladaptive: evidence from the plant Red List. *Trends in Genetics* 19: 609–614.

Vinogradov, A. E. 2004. Genome size and extinction risk in vertebrates. *Proceedings Royal Society of London B* 271: 1701–1705.

Wade, M. J., and C. J. Goodnight. 1998. Perspective: The theories of Fisher and Wright in the context of metapopulations: When nature does many small experiments. *Evolution* 52: 1537–1548.

Wade, N. 2005. Joseph Greenberg, singular linguist, dies at 85. *New York Times*, May 15, 2001.

Wakeley, J., 2003. Polymorphism and divergence for island-model species. *Genetics* 163: 411–420.

Walton, M., 2002. Apes lack the genes for speech: Discovery helps explain why humans can talk the talk. CNN. http://archives.cnn.com/2002/TECH/science/08/15/coolsc.speech/index.html

Wang, R. L., A. Stec, J. Hey, L. Lukens, and J. Doebley. 1999. The limits to selection during maize domestication. *Nature* 398: 236–239.

Wells, S., 2002. *The Journey of Man: A Genetic Odyssey*. Random House.

Weiner, J., 1994. *The Beak of the Finch*. Random House.

Will, G. F., 2005. Grand old spenders. *Washington Post*. Op-ed. November 17, 2005, Page A31.

Williams, T. N., T. W. Mwangi, D. J. Roberts, N. D. Alexander, D. J. Weatherall, S. Wambua, M. Kortok, R. W. Snow, K. Marsh, 2005. An immune basis for malaria protection by the sickle cell trait. *Public Library of Science Medicine* 2: e128–e134.

Wilgoren, J. 2005. Kansas school board approves controversial science standards. *New York Times*, November 9, 2005, p. A1.

Wise, C. A., M. Sraml, D. C. Rubinsztein, and S. Easteal. 1997. Comparative nuclear and mitochondrial genome diversity in humans and chimpanzees. *Molecular Biology and Evolution* 14: 707–716.

Wong, Y-J., and J. Hey. 2005. Divergence population genetics of chimpanzees. *Molecular Biology and Evolution* 22: 297–307.

Wright, S., 1969. *Evolution and the Genetics of Populations: Volume 2: The Theory of Gene Frequencies*. University of Chicago Press.

Wright, S., 1978. *Evolution and the Genetics of Populations: Volume 4: Variability within and among Natural Populations*. University of Chicago Press.

Yedid, G., and G. Bell. 2001. Microevolution in an electronic microcosm. *American Naturalist* 157: 465–487.

Yi, S., and J. T. Streelman. 2005. Genome size is negatively correlated with effective population size in ray-finned fish. *Trends in Genetics* 21: 643–646.

Zhou, N. N., K. F. Shortridge, E. C. J. Claas, S. L. Krauss, and R. G. Webster. 1999. Rapid evolution of H5N1 influenza viruses in chickens in Hong Kong. *Journal of Virology* 73: 3366–3374.

Index